MARS!
by Jeff Rovin

CORWIN BOOKS • LOS ANGELES

MARS!

•

Copyright © 1978 by Jeff Rovin

•

Introduction copyright © by Ben Bova

•

All rights reserved. Printed in the U.S.A.

•

10 9 8 7 6 5 4 3 2 1

•

No part of this book may be reproduced in any form without permission in writing from Corwin Books except by a newspaper or magazine reviewer who wishes to quote brief passages in connection with a review.

•

Library of Congress Cataloging in Publication Data

Rovin, Jeff.
    Mars!

    Bibliography: p.
    1. Mars (Planet)   I. Title.
QB641.R7   523.4'3   77-15949
ISBN 0-89474-011-3

•

Designed by Ward Damio and Roy Kuhlman

•

Published by
CORWIN BOOKS, a division of
PINNACLE BOOKS, INC.
2029 Century Park East
Los Angeles, California 90067

To Michael, who may one day walk on Mars

MARS!

PREFACE
THE ILIAD AND THE SPACE ODYSSEY

It is far from coincidental that humankind's first and greatest epics, Homer's *The Iliad* and *The Odyssey*, deal with adventure and exploration, or that among history's most revered personalities are Moses, Christopher Columbus, Scott of the Antarctic, and Amelia Earhart—people who braved the unknown in a quest for understanding. We realize, subconsciously if not forthrightly, that the human race matures only when it is free from fear and doubt to search the horizon for knowledge. It is an innate drive that blossoms when touched by the quality known as imagination.

[Civilization has made it incumbent upon humankind to continue its exploration of outer space. The data and tangible spin-offs of past and existing programs have bettered our lives and formed a firm foundation on which to build the future (*see chapter six*). However, of late, we have become a lethargic people, unimpressed by new developments in science or art and apathetic to discovery. Our heroes are no longer the explorers but the iconoclasts—the rebels who destroy rather than build; the cynics who echo our own rampant dissatisfaction with society. In short, we have lost our imagination, our sense of wonder.

[This book was written to help rekindle that spark, a capacity which distinguishes humankind from the other animals.

[As Ben Bova explains in his introduction, Mars is a fascinating planet, and this has certainly been a banner period in its exploration.

A pair of Viking landers has sent back the first close-up photographs and data from the Martian surface; the robot probes have changed the complexion of scientific investigation. However, glancing backward, this past year also marks the one hundredth anniversary of the awesome discoveries made by Giovanni Schiaparelli and Asaph Hall.

[Who are Schiaparelli and Hall? That's something you'll learn as this complete history of Mars unfolds. We'll also be looking at the more fanciful Yammirs, Tweel, tharks, and Shizala, and at asteroids, Argon-36, H. G. Wells, *Santa Claus Conquers the Martians*, and Valentine Michael Smith. In other words, this book examines everything even remotely connected with the Red Planet, from its dramatic birth to its viability as an abode of life to speculation on its ultimate demise.

[We may not convince you that the exploration of space is a small burden (.008 percent of the military or welfare budgets) that should be willingly shouldered by the American taxpayer. But if, as you read, you find that inbred sense of discovery tingling, as it did for those wide-eyed Greeks who listened to a blind poet spin sagas of a great Aegean adventurer—then this work will have been a success.

JEFF ROVIN
NEW HARTFORD, CONNECTICUT
APRIL, 1977

ACKNOWLEDGMENTS

The author would like to thank Jim Kukowski and Les Gaver of NASA, Yvonne Samuels of the Jet Propulsion Laboratory, A. Gary Price of the Langley Research Center, H.L. Giclas of the Lowell Observatory, Dr. Mark Chartrand of the Hayden Planetarium, Paul Toomey of the Fels Planetarium, and Ben Bova for supplying reams of scientific data. Other valuable contributions to the text were made by Geraldine Duclow of the Free Library of Philadelphia and by my editor, Andy Ettinger, and Gail Dow.

The launching of Viking I
opened an astounding era in
the exploration of the universe.

INTRODUCTION
WHY MARS?
by Ben Bova

It was Saturday, November 13, 1971, when Mariner IX went into orbit around Mars, the first human artifact to achieve an orbit around another planet.

[I had only recently assumed the position of editor of *Analog* magazine. Thanks to the lifetime of devotion of the previous editor, John W. Campbell Jr., *Analog* is the most prestigious science fiction magazine in the world, a magazine in which the science fiction stories are based solidly on real science and carefully reasoned speculation.

[So I was flattered but not surprised when I received a phone call from BBC's New York office. A woman asked if I would consent to a telephone interview from their London studios, about Mariner IX and Mars. I readily agreed—until she added:

["Now, you realize that we will also be interviewing all sorts of scientists about this. They will give us the scientific background. What we want from you is something wild . . . you know, far-out."

[I said, "We've just put a flying laboratory into orbit around another planet, more than fifty million miles away from here, and you want something *far-out?*"

[How quickly the miraculous becomes commonplace.

[The entire world watched the first Apollo astronauts step onto the Moon's dusty surface. Nobody but diehard "space nuts" watched the third or fourth Apollo missions.

[There was huge excitement about the first Viking spacecraft's landing on Mars' rock-strewn, sandy ground. But only a few weeks' later, when Viking II touched down, the excitement had dimmed.

[Space exploration is intrinsically exciting. Every person who has watched a rocket lift-off, in person or on television, has felt that excitement. But the adventure seems to wear off rather quickly, for the common TV viewer.

[Of course the Apollo program was exciting, for a while. Everyone who has ever dreamed of "flying to the Moon" felt a thrill of accomplishment as our scientists, engineers, and astronauts made the impossible come true.

[There were two main ingredients in the Apollo excitement. First, it was men going to the Moon, not merely machines. Human drama requires human participants. Second, at least in the beginning of the Apollo program, we were racing against an adversary—the Russians. The fact that they dropped out of the race, because we were so far ahead of them that they could never hope to win, robbed Apollo of some of its dramatic impact. But not much.

[However, once the "space race" died, once it became clear that the United States would have no real competition in putting the first men on the Moon, public interest in the space program dwindled to just about zero.

[Except for Mars.

[There has always been a special excitement about the Red Planet. After all, the Russians landed spacecraft on the surface of Venus long before Viking touched down on Mars, and hardly anybody noticed or cared. Venera VII landed on Venus' hellish, cloud-shrouded surface December 15, 1970. In October 1975 Venera VIII and IX not only landed on the surface, but sent back photographs of Venus' surface. Yet these failed to capture the excitement that Viking's pictures of Mars engendered.

[Even the politicians have been lenient toward Mars. While many Congressional budget cutters fulminated about NASA's expenditures on the Apollo program, Viking was seldom attacked, even though the research and hardware for Viking must have cost about two billion dollars. This amounts to less than ten dollars per person for each American citizen, a good deal less than we spend *monthly* on cigarettes or liquor. (The Apollo program, which cost about $23 billion, averages out to roughly $100 per head; still quite a bargain.)

2

[Why is Mars so exciting?

[Is it because all our lives we have heard speculations that life might exist there? Is it the thrill of taking a step across space that is measured in tens and hundreds of millions of miles, rather than the "mere" quarter-million-mile span between Earth and the Moon?

[My own feeling is that Mars is so interesting to so many people because it is the one planet in the solar system whose surface we can see.

[Venus, closest planet to Earth, is perpetually shrouded by clouds. Mercury is too small and too far away (and too close to the Sun) to be seen clearly from Earth. Jupiter and the farther giant planets are all cloud covered.

[But Mars is close enough, and its atmosphere is thin and cloudless enough, for us to see its surface easily. And what we have seen over the past century or so has made Mars the favorite world of astronomers, explorers, thinkers, and romantic dreamers.

[And, for nearly a century now, we believe we have seen evidence of life on Mars.

[This book will tell you all about the many facets of our long fascination with Mars: Why we think life may exist there, what the Viking spacecraft have told us from the surface of that dry, frozen, yet infinitely exciting world. Jeff Rovin covers all the romance and mystery of our love affair with the Red Planet, as well as explaining lucidly all the known scientific facts—and showing how many of these "facts" turned out to be errors either of information or interpretation.

[Me? I'm a science fiction writer and editor. Jeff can tell you about Mars yesterday and today. My specialty is tomorrow. Here's what tomorrow may tell us about Mars:

[Life exists on Mars. The confused and puzzling results of the Viking landers' biology experiments seem to point to activity in the Martian soil that is neither Earth-type life nor ordinary chemical reactions. Obviously, what Viking is trying to tell us is that Martian life is different from ordinary chemistry—and also different from Earth-type biology. After all, Mars is a very different world from Earth. The life that evolved there *would have to be* different from Earth-type life.

[Before the end of this century, manned (and womanned) expeditions will be sent to Mars.

[You see, the really fascinating findings of Viking are that Mars

3

possesses ample supplies of oxygen, nitrogen, carbon dioxide, and water—all the key ingredients needed to support human explorers. A Martian expedition could live on the surface of the Red Planet in much the same way that an Antarctic expedition lives on the ice near Earth's South Pole. It might even be possible for the explorers/ astronauts to grow their own food in the Martian soil, inside greenhouses, using Martian water and the minerals in the Martian soil.

[The human explorers will probe the details of Martian life, and this new understanding of how life works will add immeasurably to our store of biological and medical knowledge. Everything we know about biology and medicine, so far, is based on only one sample of life. Remember that, chemically, there is only one form of life on Earth. An amoeba is chemically the same as a movie star. On Mars, a different kind of life may show us things about ourselves that we could never discover without that "third dimensional" view that Martian life forms will give us.

[The cure for cancer may well lie in the sands of Mars.

[Even if all this proves false, even if Mars turns out to be as empty and sterile as the Moon, the Red Planet will still have served us well. For it has acted as a beacon, a lure, a goal to pull us across the millions of miles of space that separate us. Perhaps we go for adventure, or nobility of spirit, or the lure of new knowledge and new riches. The reasons are almost immaterial. The important thing is that we go.

[And in this exploring, this striving, we grow. The human race gains new skills, new wealths of knowledge and understanding, new strength and maturity.

[Perhaps this is our true cosmological mission. Even if our tiny planet is the only place in the whole vast universe where life exists—*especially* if we are alone in the universe—then it is our mission to spread ourselves throughout creation, to bring life everywhere we go, to make the entire universe our dwelling place.

[We are the life carriers. Even if Mars has never felt the touch of life in all its billions of years of existence, it has now felt our life. And we will walk on Mars, ourselves, before this century is over.

[It is written in the stars. Because men of wisdom and daring have written it there.

1

THE SOLAR SYSTEM

In the beginning, there were no atoms. Heat in excess of one trillion degrees had prevented the formation of their nuclear components, protons and neutrons, from the smaller quarks. There was only the Cluster, a matrix dense beyond imagining and within whose perimeter was every particle of matter that would one day comprise the universe. To a hypothetical observer hovering just beyond the Cluster's 620-million-mile radius, the darkness would have seemed as deep as the temperature was high. This imaginary visitor would have seen or heard nothing, felt no chill or warmth; the molecules to produce sight, sound, and sensation did not yet exist. Nor could the Cluster itself have been seen! Its gravity was absolute and prevented even the constituent photons, the fundamental units of light, from escaping.[1]

---

1 No one knows for how long the Cluster thus defined existence and nonexistence. In fact, during the 100 generations in which humankind has actively pursued knowledge, he's proven the old adage that for every answer there are two new questions—particularly where the Cluster is concerned. Scientists, for example, view it as a numbing paradox. Matter can neither be created nor destroyed, only transmuted; yet, the Cluster cannot have existed eternally. This compels us to consider religion's claim of a prime mover—until we realize that, at some time in the past, even a Creator *had to have been created!* However, if we cannot comprehend the Cluster in a *physical sense,* we can at least express it as a *relationship.* In this respect, it is tremendously significant to our study, serving as a common denominator for all matter, from galaxies to microbes. Indeed, even your contact with this page is a reunion of sorts: you and it were Cluster-mates

[Scientists cannot tell us why, but thirteen billion years ago, some inexplicable force caused a momentary imbalance in the splendid harmony of this primal atom. With a rending that was absolute, the Cluster shattered into a near-infinite number of pieces. The temperature plunged to a modest five billion degrees, while the rootless, disoriented particles were drawn together by their inherent gravity and began forming protons and neutrons. The third of the basic atomic integrants, the electron, had always existed in its indivisible state.[2]

[Three minutes speed by, and the growing cosmos drops to a balmy one billion degrees. The protons and neutrons have bonded and begun to gather, within their orbits, one or two electrons apiece; these were atoms of hydrogen and helium, the first natural elements. And, after one hundred million years of such interaction, these incomplex, primordial gases dominated the universe. By this time, of course, the gravity of matter working against matter had decelerated the issue of the initial explosion. And, left to their mutual attraction, the slowed hydrogen-helium atoms formed huge clouds or *protogalaxies*, in which millions of large gas eddies began to contract due to the play of local gravity. This same force caused the pools to attract evermore amounts of surrounding matter; eventually, the eddies or *protostars* had become so dense that they could no longer resist the magnetic or gravitational pull from the core of the protogalaxy. They fell toward this center, gathering speed and gases en route. However, due to intertia they flew *past* the core, then back again, remaining in stable orbits only when the various reactions had cancelled one another out. The shape of the framing star-family or *galaxy*—either spiral, elliptical, or irregular—was determined by this final distribution of eddies and by their gravitational effect on each other as well as on the uncoagulated gas and interstellar particles

some 14 billion years ago! So we have a valuable filiation in the Cluster, for it transforms the universe from a vast, sometimes frightening expanse of darkness, light, and alien places, to a large but fascinating *home*. And the residents of this home, whether chemical or biological, are nothing less than our kith and kin! Given this perspective, to look at Mars—or, for that matter, any other heavenly body—becomes *more* than just a pleasure or a curiosity: it is, to convoke an appropriate cliche, a God-given responsibility!

2 In 70 billion years, this spreading will have slowed, stopped, and reversed, at which time the matter of the universe will collapse inward. Only when it is reborn—as perhaps our current universe is the recycled form of some earlier system—will nonsubsistence be redefined.

The dawning of Creation, as suggested by
this Viking I photograph of a Martian sunset.

The birth of the planets as suggested by this scene from the opening moments of the motion picture *2001: A Space Odyssey* (1968).

still floating in the system. Of course, by the time the high-velocity travels of these hazy globes had ceased, their central temperatures were so intense that they succumbed to thermonuclear fusion. They became the first stars. In other, more eloquent words, it was as if some inspired presence had simply declared, "Let there be light."

[This reaction, which gave birth to the stars, was the product of a basic atomic chain. A hydrogen atom—a proton nucleus with a single electron satellite—merges with a second proton to create a heavier form of hydrogen known as *deuteron*, which itself captures a proton and is metamorphosed into a helium variant known as helium-3. This light isotope, in turn, snares a fourth proton, thus creating the standard form of helium which has an atomic weight of four. However, the total mass of four protons is slightly more than the mass of one helium nucleus, and this difference in mass is converted to energy as described by Albert Einstein in his famous equation $E=mc^2$, and which causes not only stars to burn, but our own nuclear power-plants and weapons to function. In the instance of a star, the process does not stop here: the helium becomes carbon, oxygen, and other gases, which themselves coalesce to form still heavier and more complex elements. Thus, less than a billion years after the destruction of the Cluster, the universe had its first generation of stars as we know them today. Second generation stars—like our own sun—were created from waste matter emitted by these native inhabitants of our millions upon millions of galaxies.

[The evolution of the sun, one of the less imposing of the one billion stars in the Milky Way galaxy, began over five billion years ago. However, in a subsidiary reaction which may be common throughout the universe—it is an issue hotly debated among scientists, and one we will discuss later in the book—the contraction of the solar gas cloud or *nebula* was accompanied by activity that gave rise to the planets. This chemistry consisted, at first, of atoms combining to form molecules, the particles which identify the properties of a substance, and, ultimately, small grains of matter. Once they had merged, the pebbles fell into orbit around the equator of the large central eddy, thus shaping a plate hundreds of millions of miles wide. However, the great disk was not just a random confluence of rock: the position of the various elements had been determined by the gravity of the budding sun. This pigeonholing was crucial to the make-up of the individual planets. The heavier materials, such as iron and silica, were able to weather the solar temperatures and pressures and hold close to the new star in positions roughly equivalent to the orbits of Mercury and Venus. Settling toward the center of the solar system were the less determined iron sulfides, hydrogen-oxygen bonds (water ice), carbon dioxide ice or "dry ice", and, far from the hub of the celestial family, condensed (solid) hydrogen and helium. However, this new distribution proved to be gravitationally unstable. As a result, the granular tapestry began to tear across its vast length. The planetesimals collided within their orbits and fused in random fashion, creating planetoids up to three miles wide; these, in turn, slammed one into the other, building the solar system's nine great planets and thirty-three moons. Indeed, the aftermath of these great collisions can still be seen on the cratered surfaces of Mercury, the moon . . . and Mars. Yet, the brutal birth pains were only a beginning. As these scarred *protoplanets* of diverse composition settled into the plane orbit of the dissipated disk, they began to collect up hydrogen-helium atmospheres. The two light elements, of course, had comprised over 90 percent of the original solar nebula, and the building of the sun had left enough residual gas to form thick blankets of air about each of the new planets. However, this development was followed by the first pulses of thermonuclear activity on the sun, and the planets found themselves washed by radiation and solar wind—a barrage of ionized (electrically charged) particles of light or *photons*. The force of this solar wind swept away the atmospheres of the inner planets, leaving in its wake four rocky and featureless worlds. These great spheres were further tortured with a sharp baking in solar heat; the restlessness of their indigenous elements would follow hard upon, resulting in volcanic activity and coating large surface sections of these inner worlds with great seas of lava and molten metal. Planets in the colder regions of the solar system, from Jupiter to Pluto, seem to have escaped the tumult. Their hydrogen and helium atmospheres had been able to withstand the onslaught, hardened, as they were, about the solid cores or secured by the powerful gravity of these five large and distant planets.

[So it was that nine billion years after the destruction of the magnificent seed Cluster, purged of all but its fundamental materials, each planet knuckled down to the taming of these composite parts. As we shall see, one of these worlds became a lavish observatory—and another became a target of the observatory's most romantic and scientific visions!

2

THE JOY OF MARS

One billion years have passed since the birth of the solar system, during which time the planets very methodically put their elemental houses in order. However, on at least one world—the third from the sun—the chance of widespread or lasting stasis was remote: this small, unimposing sphere had *temperament*. Shortly after its creation, heat from the natural radioactive decay of uranium and thorium, collision with meterorites, and the settling of the young planet's mass caused its nickel-iron content to melt, filter slowly downward, and create a seething, molten core. The lighter elements, such as silicon and aluminum, rose to forge a crust, while the remaining silicates formed an intermediary mass, the eighteen hundred-mile-deep mantle. Yet, the planet was still unstable and built volcanoes to release the heat newly raised by its solute core. This last maneuver had dramatic result: it not only effected a temporary equilibrium, but freed carbon dioxide, methane, nitrogen, and water vapor which had been trapped in terrestrial building blocks. In short, it provided a framework for the nursery called earth. The massive issues of carbon dioxide mixed with the calcium silicate of surface rocks to form quartz (sand), limestone, and marble; the other gases drifted slowly toward space where they were gripped and held by the earth's gravity, thus forming an atmosphere. And when the eruptions ended, and the

11

planet began to cool, condensation followed, causing the millenia-long rainstorms which resulted in our oceans.

[At the dawn of the Archaen era (3.5–1.5 billion years ago), the earth's newborn atmosphere was without its ozone layer, a blanket made of oxygen atoms joined to create the new gas, and which screens out the sun's ultraviolet radiation. Fortunately, oxygen in an unbonded state is a *product* and not a *forerunner* of life. Thus, solar rays were free to play about the earth's surface. This intense ultraviolet energy baked our rich seas, causing a synthesis of organic compounds which, in turn, joined to form larger molecules and, ultimately, nucleic acids: the basis for genetic reproduction. From this point forward, the creation of more sophisticated life forms followed hard upon, the bacteria and marine algae of the Proterozoic era (2 billion–600 million years ago) becoming the fish, insects, and small reptiles of the Paleozoic age (600 million–200 million years ago) whose evolution brought forth the awesome dinosaurs of the Mesozoic era (200 million–70 million years ago). The extinction of these huge reptiles ushered in the Cenozoic era and the advent of birds and mammals in its subsidiary Paleocene period (70 million–25 million years ago), which gave way to the cooling of the earth's climate in the Neocene period (25 million–1 million years ago), the Ice Age or Pleistocene epoch (1 million–10 thousand years ago) and, during the Pleistocene age, the coming of man.

[Until 10,000 B.C., prehistoric man lived in small nomadic tribes, harvesting wild vegetation and hunting with crude clubs and stone weapons. However, at the onset of the New Stone Age or Neolithic era (10 thousand–4 thousand B.C.), man found that he could better his chances for survival by breeding animals and planting crops for food. Thus, as far back as 8,000 B.C. and the settlement of Jericho on the north shore of the Dead Sea, humankind began to form small, self-supportive villages. There was very little communal activity in these farming towns; everyone fed, clothed, and sheltered themselves. They banded together simply for protection. However, by limiting his lateral horizons, man found that he had to depend on the earth's cooperation in order to survive. And it was this factor, oddly enough, that forced him to study the sky. Stone Age man had had little use for the heavens. He knew that the sun supplied light under which he could hunt, that the sky gave him fire in the form of lightning and water as drops of rain, and that the rising moon meant a time to seek sanctuary

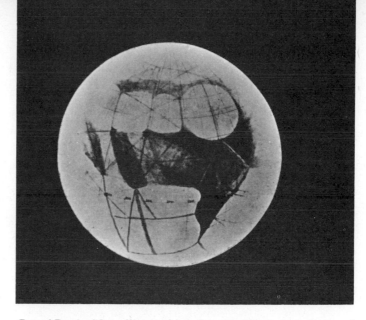

One of Percival Lowell's canal-laced maps of Mars. Lowell piqued human curiosity about Mars more than any scientist in history.

from the sharp-eyed carnivores of night. But when these Neolithic tribes formed permanent settlements, it became a matter of life and death to know what nature had in store for particular geographic locales. Unfortunately, our knowledge of specific observations made during this period is limited: none of these early peoples had a written language. Thus, it wasn't until the advent of the record-keeping Sumerians that we find a history of early astronomy.

[No one knows from where the Sumerians came. They entered—indeed, gave birth to—recorded history circa 5,000 B.C., settling with a few migrant farmers in the fertile valley of the Tigris and Euphrates Rivers—the Biblical Garden of Eden. Yet, there was a paradox to this new land of Mesopotamia (now a part of Iraq). It boasted excellent farming conditions, but otherwise offered only great stores of sand and mud. Accordingly, the Sumerians set up centers of trade in which to procure timber, rock, metals, and flint in exchange for livestock and crops. This program fostered unanticipated prosperity as well as a hurried expansion of vital resources: new farmlands were nurtured with irrigation ditches; the all-important elements were deified and honored with priests and temples; local leaders were appointed; craftsmen were trained to work the great influx of new raw material into jewelry, clothing, and tools; and the first modern cities were born. Shortly thereafter, a similar growth took place in the Nile Valley of Egypt and in Babylonia, while a quieter boom was enjoyed along the Chinese Yellow River and in the Indus Valley of India.

[To keep track of inventory, trade, and the seasons on which their crops depended, the Sumerians invented writing in the fashion of

13

One of Mariner IV's first closeup photographs of Mars. The planet's cratered surface shot down the hopes of finding fantastic civilizations on the Red Planet.

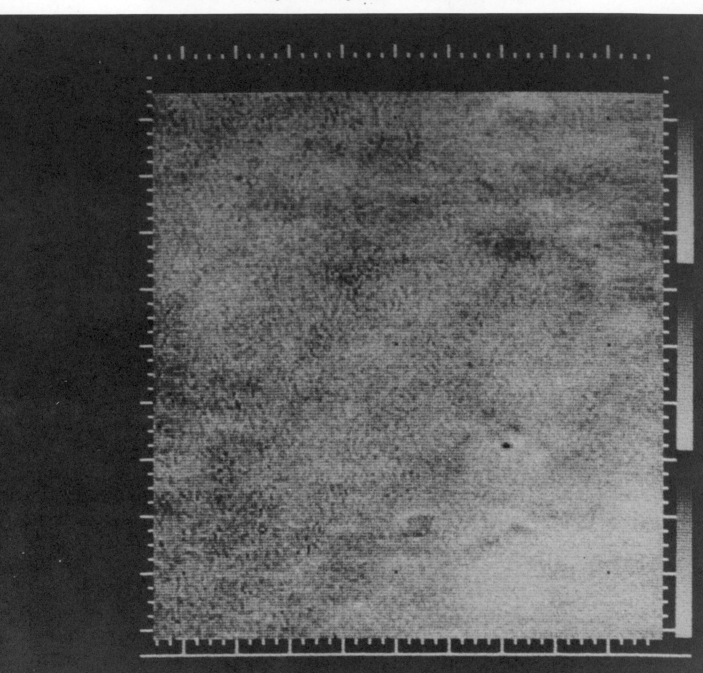

cuneiform or wedge-shaped characters. With these, they began to record the movements of the sun and the moon circa 4,000 B.C. And if we define a scientist as someone who studies and keeps track of natural phenomena, then these early Sumerians were the world's first astronomers. Ultimately, they used their observations to create a calendar based on the movements and phases of the moon, with every year broken into twelve months averaging thirty days apiece. This allowed for a coordinated harvesting of crops prior to the annual rising and flooding of rivers, the timely celebration of festivals honoring their gods, and even the regular collection of taxes. However, the Sumerians didn't stop with simple timekeeping. They had noticed that the moon regulates the tides, that a woman's fertility was mysteriously linked to a lunar timetable, and that the seasons were broken into symmetrical patterns announced by the sun's position in the sky. Thus, it logically followed that man, too, must be governed by the heavenly bodies. The moon, of course, was said to drive some men mad when its face shone full[1]; but there was more to the sky than a flaming disk by day and a lesser beacon by night. There were tiny spots of light which appeared each evening and moved about in unchanging patterns. The Sumerians further recognized that the lights in what we now know as a 10-degree swath of the *ecliptic*—that area of the sky through which the sun passes on its day to day course—repeated a cycle every twelve months. Thus, they built a celestial mosaic from these stars and called it the *zodiac*, or zone of the animals. These dozen designs, our first *constellations*, were thought to hold considerable sway over those born during the months in which the sun complimented their nighttime reign. But in those patterns were *other* lights—one of them the color of fire and referred to, by the later Egyptians, as *Horus the Red*, after their hot-tempered, falcon-headed god of war—and they roamed about the heavens in an unpredictable manner. What the Sumerians called these wanderers we cannot say, although, in any case, their interest was more superstitious than astronomical. They wanted to know how these footloose points of light affected the zodiac and their own personal destinies, and tried to chart their motions. The gauge they used was a circle

1 It has long been believed—but is as yet unproven—that the moon's gravity affects the fluid in our inner ear, the way it does the tides, thus unbalancing those individuals sensitive to the change. It was from this concept that the coming of that legendary madman the werewolf came to be linked with the full moon.

15

marked with 360 points or *degrees*, in which the sun, their foundation measurement, moved one degree per day. Each degree was further broken into sixty *minutes*, which were themselves subdivided into sixty smaller *seconds*. The resultant coordinates of the four rootless stars were rough, but they allowed astrologers to extrapolate their value in the form of horoscopes. Thus, it wasn't until the Greek civilization that astronomy moved beyond the functional and purely ceremonial levels of the Sumerians.

[For nearly twenty centuries (3,000–1,000 B.C., the Aegean landscape was dominated, first, by the Minoan and then the Mycenean civilizations centralized on the island of Crete. Then came a wave of invasions by Aryan tribes which, after an incubation period known as the Greek Dark Ages (1,000–700 B.C.), gave birth to the legendary rule of the Greeks. The Ancient Greeks were an amazing people who, beyond their contributions to the arts and government, elevated the study of nature from a survival tactic to an art. Not surprisingly, the earliest Greek astronomy was little different from that of previous civilizations. The constellations and heavenly bodies were given Hellenic names—that burning Red Planet was now Ares, after the *Greek* god of war—while the myth-makers spun tales to explain the workings of earth and sky. However, the stories in which we're interested come to us through the eighth-century poet Homer. Predating even the earliest of the philosophers—Thales of Miletus, whose seventh-century contribution to science was a theory that the earth had developed from water—Homer and his contemporary Hesiod crystalized current thought on the relation of earth to the heavens. They wrote about our world as a large plate surrounded by the great river Oceanus which, after it circled the earth, flowed back into itself. Above this flat earth was the dome of heaven which, according to Hesiod, was nine days high as the anvil falls; complimenting the celestial vault was Tartarus, the semi-circular nest of Hades. Later Greek scribes elaborated on this simple scheme by noting that the stars, planets, moon, and sun all plunged into Oceanus at the completion of their daily course, were carried about on the bosom of the water, and emerged again, unscathed, the following day. Of course, it can be argued that these ideas were little in advance of the Hindu concept that our flat earth rode through the heavens on the back of a huge turtle, or the Egyptian view of our disk-like world being carried along by two pairs of large elephants. Nor was there anything uniquely scientific in the Greek

belief that the sun, moon, and planets were set in transparent spheres, the sounding tones of which corresponded to the notes of the diatonic scale. But the early Greek thinkers *did* manage to infuse convenient mythology with the first threads of scientific observation. They not only wanted to know *when* a season, the full moon, or a solar eclipse would occur, but *why*. The process was slow at first, increasing only as their awareness of the natural world grew. And, although many of their deductions would prove to be as implausible as their tales of gods and heroes, it was first and foremost the discipline of methodology which man needed at the time. The contemporaries of Greece never passed to this level of sophistication.

[The scientific and philosophic wheels of Ancient Greece first began to turn with inspired vigor in the sixth-century B.C., thanks to a brilliant paradox named Pythagoras (582–500 B.C.). Beginning with famous mathematical theorems and proofs, Pythagoras went on to define the heavens and earth as a system of consummate harmony. It was illogical, felt Pythagoras and his legion of followers, that existence could be anything *but* beauty in the form of symmetry. Accordingly, he advanced the idea that the earth was round, for no other reason than that a sphere was perfectly balanced and a shape much desired by his school; and that the earth and the heavenly bodies all revolved around the sun, its fire giving motive and meaning to the universe. This was somewhat of a revolution in the fledgling realm of science: to date, every civilization had chauvinistically assumed that the earth was the hub of all celestial activity. Ironically, these correct conclusions, though arrived at through wrong but impressive logic, were discarded by most everyone in Greece, while they embraced a Pythagorean concept which was entirely *false*: that the orbits of the planets, whether around the sun or around the earth, were perfectly circular. Both the sense and aesthetics of this scheme seemed irrefutable. Unfortunately, observations revealed that the planets moved contrary to the dictates of circular orbits. For example, that Red Planet had a nasty habit of traveling steadily and rationally eastward along the zodiac, like the sun and the moon, after which it would slow down, come to a complete stop, and move in a westward or retrograde direction. Then it would stop again and resume its easterly course, but along a different latitude. Naturally, to explain this irregularity was a matter of some importance: Pythagoras had passed on, and his integrity was at stake! Thus, mathematicians like Eudoxos of Cnides

(408–355 B.C.) jumped to preserve the Pythagorean constructions. Charting the heavens, Eudoxos built a system wherein Ares and the other planets each revolved around the circumference of *another* circle or *epicycle* which, itself, was spinning about the earth. In other words, while the planetary trails *seemed* asymmetrical from the perspective of a terrestrial observer, they were really quite orderly. And since this system allowed for a more accurate prediction of planetary motion, it formed the basis of astronomical science for the next fifteen hundred years.

[Fortunately, the contributions of other Greek thinkers were more actual than supposed. For example, the philosopher Anaximenes (585–528 B.C.) complimented Pythagoras' spherical earth theory with his own more responsible argument that the earth *had* to be round since its shadow, cast on the moon during a lunar eclipse, was round. Also in the stead of Pythagoras, Aristarchus of Samos [310–250 (?) B.C.] worked out the mathematics of an heliocentric solar system, with the apparent motion of the stars and the erratic movement of the planets caused by the earth's simultaneous revolution around the sun and rotation about an axis. Aristarchus also tried to compute the *size* of the sun using *parallax*, the displacement of a celestial object caused by viewing it from two geographic locations. The triangle formed by the twin lines of vision and the distance between them created a solvable geometric figure. And while crude equipment caused Aristarchus' computations to be off—the scientist declared that the sun is twenty times larger than the earth when, in fact, the ratio is four hundred-to-one—his thinking was correct. Indeed, parallax worked for the Greek geographer Eratosthenes (276–195 B.C.) who used it to deduce the circumference of the earth. Planting a stick vertically in Alexandrian soil, he could tell from its shadow that the sun was 7 degrees from its zenith when, in Syene (Aswan), a similar measurement placed the sun exactly overhead. Dividing the possible discrepancy of 360 degrees by seven, and multiplying the result by the north-south distance between the two cities, he ended up with an accurate measurement of 25,000 miles. Likewise, the astronomer Hipparchus had great success with parallax when, in 130 B.C., he used it to gauge the distance from the earth to the moon. When a total solar eclipse at Hellespont measured only four-fifths total in Alexandria, he was able to compute the size of the moon's shadow and compare it with the known size of the earth's shadow during a lunar eclipse. The resultant distance to the moon was figured to be thirty times the earth's diameter (7,926 miles) or roughly 240,000 miles—the correct figure. Hipparchus, who is widely considered to be the Father of Astronomy, also made the first catalogue of stars, a complete record that went beyond just the zodiac and the one thousand stars of the remaining thirty-six Greek constellations.

[It is in the nature of things that civilizations, like anything else on earth, must ultimately atrophy and die. And by the first century A.D., Greek culture had become little more than a respected heritage. Simultaneous with this decline, however, was the growth of the Roman Empire. Founded in 800 B.C., Rome had grown from an unassuming settlement to a city, then an empire, and finally a way of life, engulfing the crumbling Greek world circa 150 B.C. Before the birth of Christ, the boundaries of this awesome republic had extended even further to include parts of Africa, Spain, Britain, and the East. But the Romans, brilliant at warfare and subjugation, did nothing for their culture other than to revere, then bastardize Greek art and religion and ignore science entirely. Oh, they Latinized the heavens—the ruddy planet Ares was now Mars, in honor of *Rome's* war god—but that was inspired by nationalism and not astronomy. Meanwhile, the Greeks, though sultry from age and the Roman sun, managed to produce one more towering figure of mathematics and astronomy, the Alexandrian theoretician Claudius Ptolemy (120–180 A.D.).[2] And among his many hypotheses—which, due to a lack of competition, became the final and thus definitive word of the illustrious Greeks—was a description of the solar system. Contrary to the Pythagorean example, Ptolemy's universe made earth the center of all heavenly rounds, although he did have the moon, Mercury, Venus, the sun, Mars, Jupiter, and Saturn follow wider but *always circular* orbits, or *deferents*. Further, in accordance with Eudoxos' construction for the defense of Pythagoras' original plan, each planet diligently traveled an epicycle as well. Unfortunately, as Eudoxos had learned, there were still anomalies of motion which Ptolemy was forced to resolve by building ever more circles within circles for greater accuracy. But because Ptolemy's geocentric system was neat and attractive, and perpetuated the honored Pythagorean scheme, it held fast through the fall of the Roman Empire (circa A.D. 400) and the rise to power of Christianity. Indeed, since it had had the obvious insight to make

2   Claudius Ptolemy is not to be confused with any previous Ptolemies, Ptolemy Soter, ruler of Egypt from 323–285 B.C., and Ptolemy Philadelphus, his son and king of Egypt from 285–247 B.C.

God's favored planet, the earth, the hub of the universe, Ptolemy's great treatise *Almagest* was adopted as the official Church position on all matters astronomical. And so it remained, even through the "Renaissance"—a period of great strides in the arts, but of frustrating stagnation in most areas of science.[3]

[History has placed the considerable weight of its admiration on the work of the man who allegedly ended the astronomical Dark Ages, Europe's first Renaissance astronomer, Nicholas Copernicus (1473–1543). Educated in Italy, the Polish-born scientist certainly deserves great praise for having reopened the case for an heliocentric solar system, and for suggesting that the earth rotates around a central axis on a daily basis. However, he advanced these ideas *not* to break from past misconceptions, but to save the clean, circular deferents and epicycles of Ptolemy. This effort was born in the face of planets which *still* refused to move where Ptolemy's geocentric plan *said* they should. Thus, harking back to the central sun construction of Pythagoras and Aristarchus, Copernicus, with absolute devotion to Ptolemaic logic, envisioned the earth and all the other planets moving about the sun on a staggering total of *forty eight different circles!* Justifying his plan—which was vaguely approximate to actual celestial movement; certainly more so than Ptolemy's arrangement—Copernicus wrote in his *Commentariolus* (1514), "Our ancestors assumed a large number of spheres for a special reason: to explain the apparent motion of the planets by the principle of regularity. For they thought it altogether absurd that a heavenly body should not always move with uniform velocity in a perfect circle." However, "having become aware of the defects" in this scheme of his forebears, Copernicus, "often considered whether there could perhaps be found a more reasonable arrangement of circles in which everything would move uniformly about its proper center, as the rule of absolute motion requires." The heliocentric system was, thus, Copernicus' solution, as well as his major contribution to science. He did scant observation on his own, drawing information from old records and the labor of uncredited aides. Therefore, we can safely say about *his* legacy what the twelfth-century Arab philosopher Ibn Rushd said about his conceptual antecedents: "Ptolemaic astronomy is nothing in so far as exis-

---

3 As it had been in Ancient Greece, the great period of artistic growth during the Renaissance was culminating with such geniuses as Michelangelo (1475–1564) and Leonardo da Vinci (1452–1519) while a comparable rebirth of science was just taking its first, tentative steps.

tence is concerned; but it is convenient for computations of the non-existent."[4]

[If Von Daniken is wrong, and there are no chariots of the gods or alien astronauts controlling our fate—a problem we will address later in this book—then we can thank whatever providence gave us Tycho Brahe. Beyond the astronomer's enthusiasm for his work, Tycho had the good foresight or great fortune to appoint Johannes Kepler as one of his assistants. Not to detract from Tycho's important astronomical studies, it was Kepler who would liberate science from centuries of Ptolemaic domination. Tycho Brahe (1546–1606) was a man with such profound reverence for the stars—and one can only respect this selfless idiosyncrasy—that he always wore his finest clothes when observing them. Perhaps his introduction to the science had something to do with it. The son of a distinguished Danish lord, Tycho was well on his way to a career in the military when a new star appeared in the constellation of Cassiopeia. This famous supernova of 1572 so transfixed the young nobleman that he turned from arms to armillae without ever looking back.[5] Of course, at that time, astronomy was not a profession deemed suitable for a grand seigneur. Thus, Tycho had to approach it in a manner befitting his station—and his considerable ego. With the support of King Frederick II of Denmark, he erected a magnificent observatory on the island of Hveen. There, he built his own instruments, published his own journals, and assumed the role of learned shepherd over a flock of apprentice astronomers.

[Tycho's work almost immediately made a shambles of many venerated Greek postulates. In particular, his initial studies of that "new star" in Cassiopeia did away with the cosmology of the Greek philosopher Aristotle (384–322 B.C.) who had deemed the heavens to be immutable. Likewise, the fact that Tycho could find no appreciable parallax for comets led him to conclude that they were at least seven times farther than the moon from the earth and not, as Aristotle had claimed, between the earth and lunar orbit. However, Tycho did join the ancient astronomers in one of their glaring misconceptions: he

4 It should be pointed out that much of what we know about Greek science and history comes to us through Hebrew and Arab translations. While Greece and her libraries were pillaged over the years, the Semites carefully kept and preserved Aegean tomes.
5 Contrast this comparative freedom to events surrounding the Crab Nebula supernova of July 4, 1054. The explosion was reported in China, Japan, and the Americas, but not in Europe. The reason: if the heavens were unchanging, there could *be* no new star! Thus, it was ignored by those who did not wish to suffer the wrath of the Church.

Anno 1666. Die 30. Martii hor: 2. n. s.

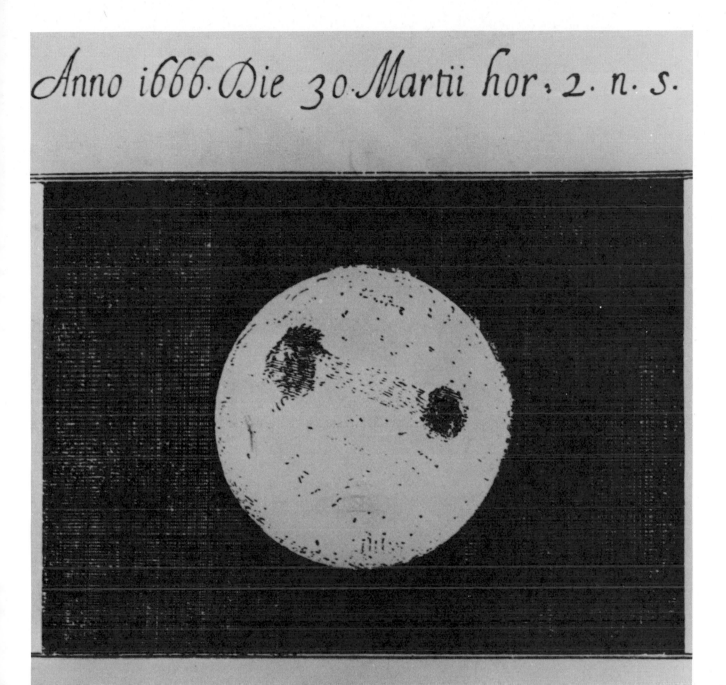

Typus Martis cum insignibus maculis Romę primùm uisis D.D. Fratribus Saluatori, ac Francisco de Serris tubo Eustachii Diuini palmorum 25, ac subinde 60. à die 2, Martii ad 30, qua die in ędibus Illmi D. Cesarii Giorii horá predicta, et ipsomet Illmo Dnó describente tub. p. 45, apparuit ut hic exprimitur inuerso modo, nigriore inter ali= as existente macula Orientali, pro situs obseruata uariatione eiusdem pla= netę circa proprium axem reuolutionis periodum indicatura, horis nempe cir= citer 13.

disavowed the Copernican solar system and put the earth back in the central position. For one thing, he wondered, "How could the fat and lazy earth be capable of motions ascribed to it by Copernicus?" For another, even the Copernican scheme did not explain planetary motion to a degree of accuracy that satisfied Tycho. Thus, he constructed a system in which, while the planets all revolved in perfectly circular orbits around the sun, the sun and the moon were *themselves* revolving around the earth! Unfortunately, Tycho's plan did not allow him to anticipate the planets with any greater precision than the unwelcome scheme of Copernicus.

[Tycho, as we have noted, had a small problem with a large ego. Spawned by his high-born background, it was no doubt fed by the scope of his work and his sovereignty at Hveen. Thus, in 1597, tired of Tycho's haughty manner, Frederick banished him from Denmark. Taking as many of his instruments and notes as he was able, the astronomer fled to Prague where he was appointed, by Emperor Rudolph II, as Imperial Mathematician of Bohemia. Gathering up a new fold of apprentice observers, Tycho resumed his celestial observations. In particular, he was devoted to making or breaking his planetary scheme by studying the planet Mars. Among all the heavenly bodies, Mars displayed patterns of movement which were in shocking contrariety to existing schemes of the solar system. Unfortunately, Tycho never realized his goal. He died in 1601, at the age of 55, passing the crusade to his most accomplished associate, Johannes Kepler.

[In a very broad sense, Tycho had set the stage for a new era in astronomy. Although he made nigh as many wrong deductions as correct ones, his legacy inspired Kepler to create a new vision of the solar system which was to become the cornerstone of modern astronomy. Like Tycho, Kepler had not started out to be an astronomer. His early training at the Swabian University of Tubingen had been with an eye toward the Protestant church. However, Kepler's masters recognized that the independent youth wasn't pious enough for a religious career. Instead, they found him a post as professor of mathematics at the Protestant Academy in Graz, where he wrote his first astronomical essay *Prodromus, sire Mysterium Cosmographicum* (1596), probing the laws that governed planetary orbits. Although it was later proven to be inaccurate, this work brought Kepler to the attention of Tycho Brahe. And when the call came from Prague in January of 1600, he jumped at the chance of working with the renowned scientist. Indeed, as Kepler was later to remark, it had been fortunate "that I arrived just at the time when Longomontanus [a Tycho aide] was occupied with Mars. For Mars alone enables us to penetrate the secrets of astronomy which otherwise would remain forever hidden from us." Unfortunately, Kepler's association with Tycho lasted less than two years, ended by the master's death. Kepler was named Imperial Mathematician in his stead, and picked up where Tycho had left off on Mars. However, he used his distinguished mentor's data not only to redetermine the planet's movements, but to satisfy a nagging conviction that the heliocentric system of Copernicus might indeed be correct. Thus, with Tycho's detailed records of Mars on hand, as well as such mathematical tools as the Martian line of apsides—the direction of a line connecting Mars at its nearest and farthest from the sun—he slaved over the problem for five years. And what he discovered was that Copernicus had only been *partially* right. As Kepler revealed in chapter forty four of his massive, fifty-chapter work *Astronomia Nova* (1609), not only was it clear that the earth was a planet and revolved around the sun like the other planets, but that "Mars' path is not a circle; it curves inward on both sides and outward again at the opposite end. Such a curve is called an oval. The orbit is, therefore, not a circle but an oval figure." Kepler would later qualify this conclusion by noting that the orbits of Mars and the other planets are more properly ellipses.

[As we have seen, the pursuit of all scientific knowledge—then as now—depends upon the favor of a government patron. Thus, after the death of Rudolph in 1612, a fickle court denied funds for Kepler's work and he had to accept the post of Provincial Mathematician at Linz in Upper Austria. He spent the next fifteen years there, working on his famous planetary laws and tables, after which he was again forced to move by the encroaching Thirty Years' War. Wandering about Europe for the next two years, he finally arrived in Regensburg, Bavaria. However, his already frail constitution had been drained by a trip on horseback through inclement November weather; the 59-year-old Kepler died after a brief illness.

[Unlike the astronomers who preceeded him, Kepler was a man who had humbly acknowledged his sources and assistants; a man whose writings, unlike those of Copernicus and later, Galileo, were romantic, readable, and filled with the exuberance he felt for his work. Thus, we must feel some sorrow for the relative obscurity in which he spent his final years, as well as his failing eyesight which had prevented him from enjoying an astronomical landmark that occurred shortly before his death.

[A young man whom we know today only as an apprentice to Dutch spectacles-maker Hans Lippershey (1570–1619) was passing the time in his master's absence. Finished with his chores, and having complete run of the shop, he began to peer through various lenses. Eventually, he put one in front of another and found that the combination made faraway objects appear to be near. The young tradesman presented his discovery to Lippershey, who was immediately struck by its potential. Holland was at war with Spain, and all that stood between the Dutch and defeat was their powerful navy. The optician realized of what enormous value these lenses would be if placed in a tube that enabled seamen to spot and spy on a still-distant enemy. The government, too, was impressed, and agreed to underwrite the manufacture of these instruments. However, they asked that Lippershey produce the "lookers" two abreast for stereo vision. Although the war ended before either the telescope—a term coined in 1612 by the Greek mathematician Ioannes Dimisiani and meaning "to see at a distance"—or binoculars, could be applied to the Dutch cause, they were not destined to disappear. And while we can only guess at the great joy they would have brought Kepler, we can content ourselves with their inventive use in the hands of Galileo Galilei (1564–1642), the first man to turn a telescope skyward.

[Galileo was born in Pisa, the son of musician Vincenzio Galilei. Educated at a Jesuit school near Florence and, later, at the University of Pisa, Galileo made discoveries in physics which early on earned him an impressive reputation. These led to an appointment as Professor of Mathematics at Padua, Venice, a position the 28-year-old scientist was to hold for 18 years. This academic period was distinguished primarily by lectures and the invention of handy, if unastounding, mechanical devices. But there was freedom in Venice, something the man from Pisa did not have at home. By the very nature of his experiments, Galileo had found himself none beloved by the powerful Church hierarchy in Pisa. For, despite the findings of Copernicus, Brahe, and Kepler, Christendom refused to release its hold on Greek science and philosophy. And they were not about to change for Galileo. The Venetians, however, were not as concerned with strict Church dogma, and Galileo was free to perform his experiments without fear of persecution or denunciation. However, the conflict which would ultimately find him on trial before the infamous Inquisi-

28

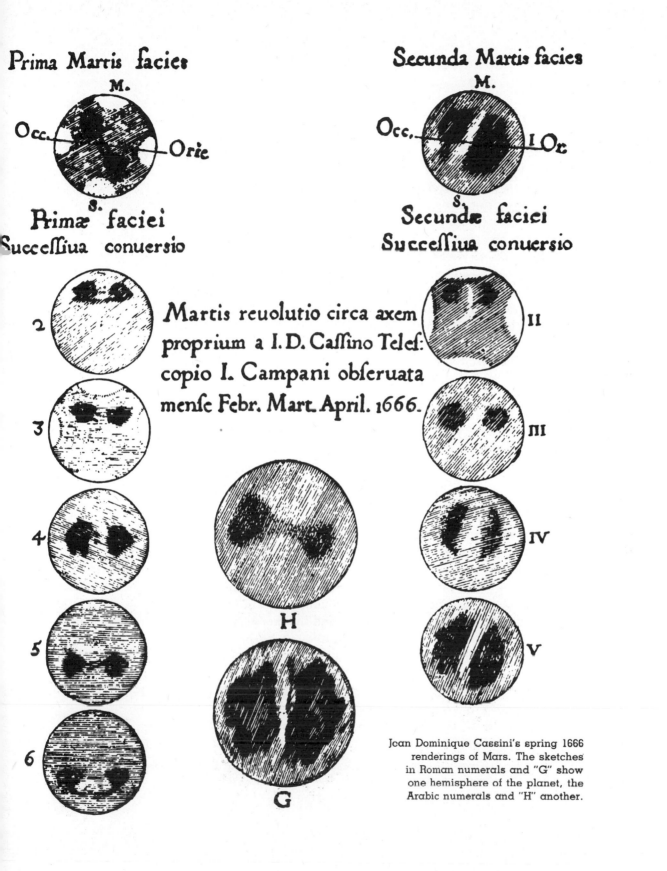

Prima Martis facies

M.

Occ. — Orie

S.

Primæ faciei
Succeffiua conuersio

Secunda Martis facies

M.

Occ. — I Or

S.

Secundæ faciei
Succeffiua conuersio

Martis reuolutio circa axem
proprium a I. D. Caffino Telef:
copio I. Campani obferuata
menfe Febr. Mart. April. 1666.

2

3

4

5

6

II

III

IV

V

H

G

Jean Dominique Cassini's spring 1666
renderings of Mars. The sketches
in Roman numerals and "G" show
one hemisphere of the planet, the
Arabic numerals and "H" another.

tion was about to begin. Toward the start of his last year at Padua—and from what source we are not quite certain—Galileo heard about a magnificent invention that was being referred to as everything from a "perspective glass" to an "optic tube." Intrigued, Galileo decided to build his own "cannocchiales" using a plano-convex and a plano-concave (eyepiece) lens. Given his superior mechanical and optical skills, there is no doubt but that this was the finest telescope produced to date.

[When the 45-year-old scientist cast his instrument toward the heavens and boosted his vision to three times that of the unaided eye, the sights he saw were truly staggering. But before we look at these, there is some insight to be gained by exploring the man and the moment. What did Galileo *feel* on that historic summer night? Did he sense the weight of history about him as, alone in his quiet Padua workshop, he gazed at sights that no man had ever before seen? Was he, himself, awed by his discoveries? Alas no, for Galileo was not an astronomer at heart. He looked at the heavens simply because they were *there* and it was something to *do* with his new toy. Thus—and regrettably—his observations were superficial and inaccurate; certainly they were not the records or revelations they could have been. Nor did the duration of Galileo's celestial studies indicate a particular interest in the heavens. By 1611, after two very productive years, he had put aside his telescope and would return to it only on rare occasions during the next thirty years. But Galileo's work, though falling far short of its potential, was still a quantum leap in man's understanding of the heavens. For example, he was the first to realize that the moon's surface was not smooth but *textured* with pits and mountains; he discovered four satellites in orbit around Jupiter and, most importantly, determined that the planets were *not* points of light like the stars, but globes like the earth and the moon. He was the first to observe the phases of Venus and, although he never saw either polar caps or surface markings, found that the flame-colored Mars, too, had a phase—which occurs at quadrature[6] and obscures 12 percent of the planet's surface. Finally, he was shocked by the sight of arms apparently growing from the sides of Saturn! This last puzzle was one that he would never unravel. Indeed, not until 1655, thirteen years after Galileo's death, did the Dutch astronomer Christian

6   Quadrature: the position in which a planet is said to be when a right angle is formed by the distance between the planet and the sun, and the sun and the earth.

Huygens determine that these protuberances were not limbs or illusions but rings. However, the story of Galileo does not end in 1611 with his faltering interest in telescopes.

[Thusfar, we have made little mention of the danger which the all-powerful Church represented to scientists like Tycho, Kepler, and Galileo. The clergy would not forsake the Greek thinkers and tolerated only that scientific work which did not create any epistemological waves. And Galileo was about to cause *just* such a stir. Not unlike his predecessor Tycho Brahe, Galileo had an irritable egomania and subtle, biting sarcasm. Thus, while he could safely refer to the work of Kepler as so much "astrological superstition," and play government seats one against another for the right to use his precision-tooled telescopes as weapons of war, he could not taunt the Church and get away with it. Lured by a sinecure that would allow him to abandon teaching and conduct research fulltime, the scientist moved to Florence in 1610. Thus, Galileo no longer had the protection of a tolerant regime when, through later telescopic study of sunspots, he was able to determine that the sun rotated about a central axis. This lent credence to the Copernican idea that our earth enjoyed diurnal rotation—a concept severely at odds with the constructions of Ptolemy. And, although Copernicus had taken care to have *his* most dangerous theories published posthumously, Galileo saw fit to bring forth *Dialogues on the Two Chief World Systems* in 1632. This work, which proclaimed to argue fairly for both the Ptolemaic and Copernican schemes, played its pro-Greek arguments with tongue so firmly in cheek that an that an Hellenic supporter could not help but realize that he had been mocked. Certainly one of Galileo's benefactors, Pope Urban VIII, got the point and, the following year, reluctantly brought the esteemed mathematician before the Inquisition. The charge: "Having held and taught the Copernican doctrine." The penalty: torture and imprisonment, if Galileo chose to stand by his work. Deciding that there was little to be gained by upholding principle in the face of ignorance, the scientist admitted that *Dialogues* was riddled with inaccuracies, and his punishment—merciful, by Inquisition standards—was to be placed under house arrest for the remainder of his days. But these last desperate attempts by the clergy to preserve their scientific heritage were straw huts in a mounting storm. By 1650, the noble Greeks had fallen for a second time, victims of a happy contagion called education. New theories were offered on the ruins of the old, accompanied by new tools and new observations, with the result that for the first time since the founding of Jericho, the scope and nature of the universe began to unfold in a truer perspective. Kepler, for all his insight, had underestimated the size of the solar system by 700 percent! And Galileo realized how far from earth the stars *really* were when they appeared in his telescope no larger than when viewed with the naked eye. Clearly, this was a period of great enlightenment, but it was only just beginning. The active pursuit of astronomical data—and not just its interpretation—became a policy of sophisticated European courts: among the first tangible results of this new and aggressive mood were the establishment of the Royal Observatories in Paris (1667) and in Greenwich, England (1675). But the most promising development of this era was the founding of the French Academy in 1666. A pet project of Louis XIV, the job of the French Academy was to gather and assimilate new scientific knowledge and, wherever possible, promote its general application throughout France. Unlike the amateur scientists of the Royal Society—their newly-formed and aristocratic counterpart in England—the men of the French Academy were working professionals assigned to specific tasks by the king. And one of the scientific body's first major undertakings was to accurately ascertain the distance of Mars from the earth at its nearest opposition—a time when the earth is between the Red Planet and the sun. This was accomplished by building a triangle with an observer in France, another in French Guiana, and the planet itself forming the vertices. The computed results placed Mars at a distance of 32 million miles; the actual figure was 35 million, representing an impressive display of mathematic acuity. This determination, incidentally, was the first step in measuring the solar system, a job accomplished with further triangulations using Mars, the earth, and the sun.

[While the French Academy achieved its goal, and engaged itself in other very beneficial studies, it was an Englishman who had the most *dramatic* effect on seventeenth-century science. Best known for having discovered the law of universal gravitation, Sir Isaac Newton (1642–1727) was also responsible for the first comprehensive exploration of the spectrum—the component colors of sunlight—a program he conducted in 1666. Working with a prism and solar rays, he laid to rest the misconceptions which had this rainbowlike band as a phenomenon of the atmosphere or of the prism itself. However—and somewhat

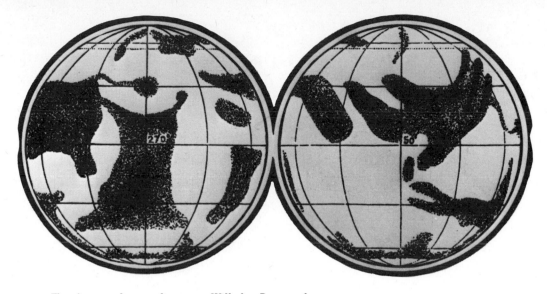

The famous lunar observers Wilhelm Beer and
J. H. von Madler drew this map of Mars circa 1831.

inexplicably—the usually thorough Newton either did not see or paid
scant attention to an important element of the spectrum, the black
bands which streak its entire length. In fact, it wasn't until 1814 that
anyone gave these stripes a second look, in this case, the German
optician Joseph Fraunhofer. Yet, it was 1859 before physicists Robert
Bunsen and G.R. Kirchhoff realized that the so-called Fraunhofer
Lines in the spectrum of iron exactly matched hundreds of the dark
bands to be found in the spectrum of sunlight. Their conclusion?
Logically enough, that there is iron present in the atmosphere of the
sun. This discovery led to the science of *spectrometry*, or the study of
elements that comprise the atmospheres of celestial bodies. It is a
development we will examine more closely in a later section.

[To this point in history, few astronomers studied the planet Mars
with any greater enthusiasm than they did the other celestial bodies.
This is quite understandable: except for its striking color, the Red
Planet posed less of a puzzle to scientists than the cratered moon or
the ringed Saturn. However, the eighteenth century would change all of
that, bringing Mars to the fore where it has remained ever since. And
the man responsible for this notoriety was, as so many of his predecessors, someone who came to astronomy quite by chance.

[William Herschel (1738–1822) was born in Hanover Germany and,
like Galileo, was the son of a musician. After serving in the military,
William followed the footsteps of his father, Isaac, and pursued a
career in music. The decision was sound: in 1767 he was awarded the
coveted post of organist at the Octagon Chapel in Bath, England. At
about this time he also happened upon a book of astronomy, and took
to the evening sky as a relaxing hobby. However, Herschel was
quickly hypnotized by this nocturnal pursuit and was soon behind

33

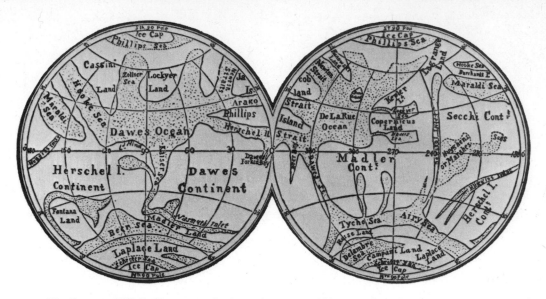

The Reverend W. R. Dawes made these drawings of Mars in 1864–65. The drawing on the left illustrates Sinus Sabaeus, Margaritifer Sinus, and Aurorae Sinus. The sketch on the right is of Mare Cimmerium and Syrtis Major.

homemade telescopes of a superior quality to most professionally built models. They had to be: in 1781, Herschel discovered the planet Uranus. Inspired by his success and continued fascination, Herschel gave up music for good. His first problem, of course, was to somehow earn a living. He had officially been named the King's Astronomer and managed to coax a stipend from George III—the selfsame tyrant of American Revolution fame—but it was a modest sum; fortunately, Herschel was also able to turn a crusade into a small business. Realizing that the success of his work depended upon the precision nature of his telescopes, Herschel set about building these after his own designs. And, as good luck and a lack of competition would have it, he made ends meet by manufacturing and selling duplicate telescopes to fellow astronomers.

[Although Herschel's primary goal was to completely and accurately rechart the heavens—including his own observations of stars, double stars, and nebulae—he also made the most thorough study of Mars ever undertaken. His findings were quite remarkable. As he revealed to an audience at the Royal Society in 1784, "It appears that this planet is not without considerable atmosphere; for besides the permanent spots on its surface, I have often noticed occasional changes of partial bright belts; and also once a darkish one, in a pretty high latitude. These alterations we can hardly ascribe to any other cause than the variable disposition of clouds and vapors floating in the atmosphere of the planet." He went on to say that, "Mars had a considerable but modest atmosphere, *so that its inhabitants probably enjoy a situation in many respects similar to our own.*"[7] Almost as an afterthought, Herschel had added a new ingredient to

7 Italics are the author's own.

34

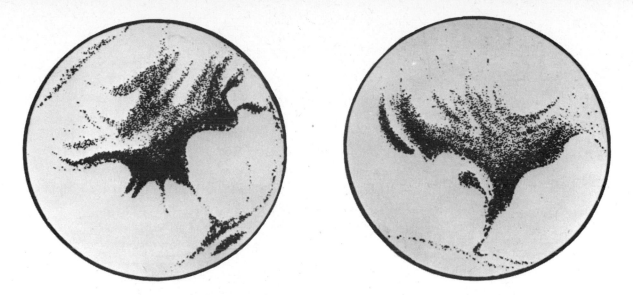

Richard A. Proctor's 1867 map of Mars. The regions he saw were named in honor of other astronomers; most of these labels were changed in standardized maps.

man's study of the stars: the concept of extraterrestrial life! Indeed, Herschel believed that all the planets, as well as the sun and the moon, were inhabited. And why not? If life forms on earth could react in harmony to the environment, might not the same situation exist on other worlds? But the thought of it—like the thought of some day *visiting* the planets—was talk for the drawing room and nothing more. In addition to these theories about the atmosphere, Herschel determined that the period of Mars' rotation was 24 hours, 37 minutes; discovered the polar caps, which he decided were snow; and divided the dark and bright surface masses of Mars into sea and land, respectively.

[Given the remarkable scope of Herschel's work, it should come as no surprise that with his death in 1822, an era of great discovery came to an end. Successors to the legendary astronomer built slowly and carefully upon his observations; those who explored the Red Planet via telescope did so with the intent of mapping the world—and little more. Thus, the names of cartographers like Beer, Maedler, Flammarion, Proctor, Green, Dawes, and Serra pass by without great significance, while their drawings, today, seem little more than quaint, sketchy relics. Indeed, only one notable achievement would highlight the post-Herschel era: the realization of a prediction made by Johannes Kepler. Kepler had reasoned that since Mercury and Venus had no moons, earth one, and Jupiter and Saturn several apiece, then Mars must have two satellites. While this line of reasoning was somewhat unscientific, Kepler's speculation proved to be correct. Unfortunately, the moons were not discovered until the sesquicentennial of Kepler's death. Working at the United States Naval Observatory in Washington, D.C., astronomer Asaph Hall (1829–1907) was determined to find these moons at the time of Mars' closest approach to earth, which occurred during the summer of 1877. The obstacles were formidable. Any dim star in the vicinity of Mars might be mistaken for a moon. The distinction could be made only after several nights of observation: if the object did not follow Mars about the sky, then it was a star. Asteroids presented another problem. The orbits of Mars and Jupiter form the outer boundaries of the asteroid belt, a collection of huge planetoids in solar orbit. Like the stars, they had to be carefully watched, charted, and eliminated. And so, in the late summer, Hall began scanning the skies around the Red Planet. The *modus operandi* was to start at some distance from Mars and systematically work toward the surface. However, as Hall's observations carried him near the planet, sunlight reflected from its surface first diluted, then all but washed away his visibility. The astronomer became terribly frustrated and gave up his quest. Then, as legend has it—and historians are fond of legend; it turns an otherwise stoic tale into high drama without radically altering the truth—his wife urged him to return for a final night's observation. Reluctantly, Hall did as she suggested and, on August 11, *did* detect a small object in Mars orbit. However, as if on some devilish cue, clouds moved in and it was five exasperating days before the astronomer could get another look. When he was finally able to observe the small body, he found that it was, indeed, a satellite; the next day he discovered a second point of light circling Mars. In keeping with tradition, the privilege of naming the moons fell to their discoverer: he called them Phobos and Deimos—Fear and Terror, after two sons of the Roman war god.

[It is ironic that this event, which caps the initial age of celestial discovery, should carry a symbolic reference to the origins of astronomy. Not only do these names pay respectful tribute to the ancient gods, but they are a reminder of man's earliest approach to the heavens. We had progressed from a time of fear and ignorance, when stars were thought to be the glistening skin of a huge dragon who swallowed the sun at night, to a glorious era of rational if misguided thought, and finally to a period of scientific repression followed by an explosion of observation, learning, and discovery. What had begun as superstitious awe was now scientific awe—and we shall examine the difference between the two as manifest in man's mythology, both old and new, about the planet Mars.

3

MARS
AND THE FANTASIST

To paraphrase Shakespeare, a planet by any other name is still that planet. Yet, for so long has Mars been affiliated with one or another god of war that it bears their mantle by association. Not that Mars is unique in this respect: Venus, for example, is identified with the Goddess of Love and thus serves as the silvery "star" of lovers. However, we must admit that it is a pleasanter lot to be known as a planet of love than as the seat of barbarism and cold superiority in our solar system! We know, of course, how this grim alliance came about. But what has it done to the planet? And how does it relate to the Mars of later centuries? Let's examine this legacy as it was manifest in the lives and customs of our ancestors.

[In both Greek and Roman mythology, the God of War was one of two legitimate children born to Jupiter and his sister Juno (the Greek Zeus and Hera), the King and Queen of the gods. But where his fellow immortals caused or partook in combat only when faced with no alternative, Mars thoroughly *enjoyed* the sight of maimed and bleeding flesh. Not surprisingly, the temperate Greeks represented Ares as being somewhat of a coward. Wounded by a mortal during the Trojan War, the god resolved thenceforth to watch all bloodletting from the sidelines. Conversely, to the militaristic Romans, Mars personified the very *spirit* of their illustrious empire. Indeed, it was Mars who fathered Remus and Romulus, the legendary founders of Rome. Accordingly, the god was celebrated with only slightly less reverence than Jupiter himself.

[The name Mars comes from the Oscan *Mamers*, the supreme deity of an early Roman satellite. Fused with the adopted Greek Ares, this inherently mighty icon became the standard of the Roman conquerors. The very word for militancy, martial, was coined in Mars' honor; a widely feared band of mercenaries during the First Punic War (264–241 B.C.) proudly referred to themselves as the *Mamertini*—the Sons of Mars. And as time went on, additional glories were lavished upon the deity. Not only was the fiery wanderer of the nighttime sky named after him, but following sun-day and moon-day in the Roman week came Mars-day.[1] The first month of the Roman year was *Martius*, and its first day, New Year's Day, was called *Feriae Marti*. On a somewhat grander scale, two festivals were held annually in Mars' honor: the springtime preparation for war, or the *Martius Quinquatrus*, which ushered in the *Tubilustrium*, the purification of the war trumpets; and, in the fall, the *Equirria*, or two-horse chariot races, which Mars shared with an agricultural god.[2] After the contests, one of the winning horses was decapitated and decked with cakes as a sacrifice to Mars. However, inconsistent with the importance of the god was the frequency of his temples in and about Rome. Even through the greatest period of Roman domination (500–78 B.C.), Mars had only two altars, one at the Campus Marius, a military and athletic training ground, and the other at Porta Capena, far beyond Rome proper. The rationale for this disposition was that while Mars should be honored and his credenda pursued, both were best kept, symbolically, away from Rome proper. However, this situation was soon to change. With the assassination of Julius Caesar in 44 B.C. and the ascension of Augustus, Mars became Mars Ultor, avenger of the slain leader and personal guardian to the new emperor. Even the sacred spears of Mars—potent weapons which, when held aloft by a special consul chanting *Mars Vigilia! Mars Arise!*, assured Rome of victory in battle—were moved from without the city to the regia, the

1   In modern times, it is still the Day of Mars. A clear genesis can be seen in the Italian *Martedi* and the French *Mardi*. *Tuesday*, the English word, has the same meaning, although it is based on the Anglo-Saxon *Tiwes daeg*.
2   Those elements of Mars which did not come from Mamers may, in fact, have begun with the agricultural deity Silvanus. This god of the fields and flocks became a warlike icon because he was the supreme deity of a warlike people; needed a warrior's attributes to protect the all-important crops; and/or was simply transmogrified by the Romans.

emperor's house in Rome. However, Mars was brought literally to the zenith of his grandeur with Augustus' dedication, in 2 B.C., of the god's new and magnificent shrine at Philippi. Unfortunately, soon thereafter, Rome as well as the war god fell before Jesus—ironically, the Prince of Peace—and the great honors faded . . . save one.

[In the wake of the Greek and Roman mythologies about Mars and stories using the sun, the moon, and the stars as a playground for their characters, the literature of the later European empires was singularly sober. *Eventually*, authors would write about the Red Planet, investing it with the swagger and dignity of its namesake. In the meantime, medieval writers dared not even mention the cosmos: only God and His minions were permitted to broach the heavens! Thus, while Cervantes was busy creating his masterpiece about a mad knight, and Shakespeare was staging plays about a mad Dane and an ambitious thane, society's imagination was kept only peripherally alive by the work of the scientists like Tycho and Kepler. Then came the discoveries of Galileo and the end of the Inquisition, and writers began to draw tentatively upon astronomy for material. The first scribe so emboldened was Cyrano de Bergerac (1619–1655), the French swordsman and adventurer immortalized in a play by Edmond Rostand. Although Cyrano traveled only to the sun and the moon—via rockets, gunpowder, and bottled dew—he was the first modern-era character to leave the earth in a work of what would one day be known as science fiction. Other little-known personalities followed in his stead, but it wasn't until the first half of the eighteenth century that the planet Mars appeared in a work of fiction. However, the brief reference was *far* from fanciful, and its origin has kept scientists and space buffs arguing for over a century.

[Oddly enough, the passage occurs in Jonathan Swift's *Gulliver's Travels*. Published in 1726, this fantasy satire follows Dr. Lemuel Gulliver as he journeys through European government as caricatured in bizarre, fairytale lands. While the first two adventures, set in Lilliput and Brobdingnag, remain the narrative's most enduring chapters, the third of Gulliver's eight voyages takes him to the fascinating flying island of Laputa. And, being concerned with the physics of flight, the Laputians naturally had a number of navigators and scientists on hand. According to Gulliver, many of these men spent "the greatest part of their lives in observing the celestial bodies, which they do by the assistance of glasses far excelling ours. This advantage hath enabled them to extend their discoveries much farther than our astronomers in Europe. They have made a catalogue of ten thousand fixed stars, whereas the largest of ours do not contain above one third part of that number." Here, however, a century and a half before Asaph Hall, Swift dropped his bombshell. Gulliver continues:

> They have likewise discovered two lesser stars, or satellites, which revolve about Mars; whereof the innermost is distant from the centre of the primary planet exactly three of the diameters, and the outermost five; the former revolves in the space of ten hours, and the latter in twenty-one and an half; so that the squares of their periodical times are very near in the same proportion with the cubes of their distance from the centre of Mars; which evidently shows them to be governed by the same law of gravitation, that influences the other heavenly bodies.

[The obvious question is how did Swift know about these moons? Erich von Daniken, in his painfully spurious *Chariots of the Gods* (see chapter seven) insists we do not know—but intimates that Martians visiting prehistoric earth may have somehow shuttled the information down through the centuries. Other dubious authorities suggest that Swift was, himself, a Martian, or that he had at least visited the planet via flying saucer. Then there was the quite romantic notion that Swift and a lover had spotted the moons during the opposition of 1688, but had kept the discovery a secret since the satellites were "their private stars." Only with the woman's death did Swift reveal their presence. Interesting though these theories may be, however, the explanation is disappointingly simple. Kepler, as we have noted, had anticipated the two moons, and the mathematics quoted by Swift were common knowledge. Thus, there is no reason to assume that Swift made anything more than an educated guess. For *despite* his attention to known astronomical data, the *Gulliver* extract ultimately reveals its own debt to chance: Mars would have to be six times its actual size in order to accommodate the statistics described by the author. So the "mystery" is a fancy union of calculation and intuition . . . and nothing more.

[Mars' next appearance in the pages of fiction did not come for another one hundred and fifty-four years. During the interim, works of science fiction featured such alien beings as the eight-mile-tall resi-

dent of the star Sirius, who visited our world in Voltaire's *Micromegas* (1752) and, in the mid-nineteenth century, Jules Verne's internationally popular trips to the bottom of the sea, to a comet, and to the moon. But Verne, like de Bergerac, avoided the Red Planet in favor of more dramatic worlds. Then, in the same year that Asaph Hall discovered Phobos and Deimos, something happened to make Mars the most exciting of all the celestial bodies. We found canals crisscrossing its face, and neither outer space fantasy or astronomy would ever again be quite the same.

[Treasured dreams die hard, and the ninety-year-old vision of canals was one such dream. It was reluctantly surrendered in 1965 when the United States' Mariner IV space probe flew past Mars at a distance of 6,200 miles. The craft's cameras gave us our first close-up view of the planet, presenting a surface marked not with canals, but with craters like those on the moon. What a glorious history ended with this revelation! It all began in 1877, with the tenacious study of Mars at close opposition by Italian astronomer Giovanni Schiaparelli (1835–1910). While Schiaparelli's work covered every point in the solar system, his particular interest lay in lines which laced the surface of Mars. These had been previously recorded but were ignored due to their indistinguishability. After charting seventy nine of the interconnecting threads, which he dubbed *canali*—channels or canals, as they were translated—Schiaparelli proposed a monumental theory: that the canali "present an indescribable simplicity and symmetry *that cannot possibly be the work of chance.*"[3] Herschel's musings about the existence of life on other worlds might not have been so far afield after all! Not that the scientific community greeted Schiaparelli's announcement with praise or unanimity: everyone looked at Mars in 1877, but only the Italian astronomer saw canals! Of course, nowhere in the world was visibility as fine as it was at Schiaparelli's Brera Observatory in Milan. The atmosphere was unusually still and allowed for crisp, undistorted views of space. Still, the sophisticated scientists of this era wanted hard fact, not Herschelesque fantasy or the excuse of shifting air blankets. If, they reasoned, there really were such markings, then other astronomers would have seen them! Indeed, even Schiaparelli was unable to spot the canals for another two years. And while a handful of independent sightings in 1886, 1887, and 1892 corroborated the charts of Schiaparelli, those who couldn't see them first hand remained unconvinced. That's when Percival

3 Italics are the author's own.

Lowell entered the drama, an American astronomer who not only saw the canals but *photographed* them.

[In the years since his death by apoplexy in 1916, Lowell has become something of a legend. The last of the Romantic astronomers—stargazers who searched the sky with as much a sense of poetry as science—his books on the subject are both artful and illuminating. However, Lowell is best remembered for having "clarified" the mystery of the canals using an equal measure of observation and imagination. Born in 1855 to a distinguished Boston family, Lowell received his education at Miss Fette's dame school, Noble's school and, finally, at Harvard, from where he graduated at the age of 21 with a degree in mathematics. Lowell went directly into business, helping to manage his family's cotton mills and utility companies. He subsequently invested several years as a diplomat, traveling throughout the Far East and recording his experiences in a number of fascinating books. Then, at the age of 38, after a long but platonic love affair with astronomy, Lowell learned about the controversy raging over Schiaparelli's interpretation of the Martian canals. Enthralled by these findings, and moved by Schiaparelli's failing eyesight, Lowell decided to continue the exploration of Mars in his stead. However, the wealthy Lowell, like Tycho before him, never did anything halfway. Locating that portion of the United States over which the atmosphere was the steadiest—in Flagstaff, Arizona Territory—Lowell built an observatory for the observation of Mars during its 1893–94 opposition. He was assisted in this endeavor by the noted Harvard astronomer W.H. Pickering, who was also an expert in the movement of earth's atmosphere. The domed observatory was quickly erected, stocked, and ready for use by mid-1894, although Lowell's studies of Mars had been going on during its construction. Of these first months at Flagstaff, Lowell wrote in *The Annals of the Lowell Observatory (Volume One)*, "the chief results obtained were . . . the detection of the physical characteristics of the planet Mars to a degree of completeness sufficient to permit of the forming of a general theory of its condition, revealing beyond reasonable doubt first its general habitability and second, its particular habitation at the present moment by some form of local intelligence." However, before pursuing Lowell's ideas about life on Mars, let's look at the data he either gathered or confirmed, a veritable summation of our knowledge at the turn of the century.

[Using new equipment and his superb mathematical training,

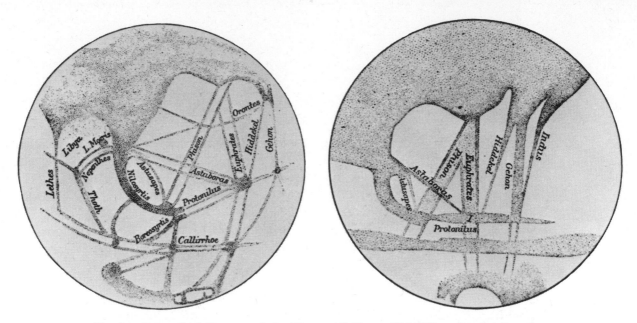

The drawing on the left was made by Giovanni Schiaparelli in Milan, the sketch on the right by Henri Josef Perrotin in Nice. Both were rendered on June 4, 1888—Perrotin's one hour after that of Schiaparelli.

Lowell determined that Mars required 686.98 earth-days to complete a single orbit of the sun; that the distance to Mars during opposition, at its closest approach, or *perihelion*, was 35,050,000 miles and, at its greatest distance, or *aphelion*, was 61,000,000 miles; and that these cycles of opposition repeated themselves every fifteen years. He further reasoned that at its greatest distance from the earth—on the opposite side of the sun, for example—Mars could be as much as 247,900,000 miles away. As to its relationship with the sun, Lowell placed Mars' elliptical orbit at 129,500,000 miles at perihelion and 154,500,000 miles at aphelion. Remeasuring the planet's diameter, he arrived at a figure of 4,215 miles or slightly less than one half the earth's diameter (he was twenty nine miles shy of the actual figure); he determined its mass to be .094 that of the earth (the correct mass is .108); and found its gravity to equal just over one-third (38 percent) our terrestrial force. Lowell also calculated the distance of Phobos and Deimos from Mars as 5,800 and 14,600 miles and orbiting the planet in 7 hours, 39 minutes and 30 hours, 18 minutes, respectively, and timed the Martian day at 24 hours, 37 minutes, 22.7 seconds, or forty minutes longer than the earth's 23 hours, 56 minutes. These figures are consistent with what we know today. However, while these basic readings, along with explorations of the other planets,[4] were being under-

4 In addition to the study of Mars, Lowell predicted the existence of Pluto based on eccentricities exhibited by the orbit of Uranus. Based on Lowell's research, Clyde W. Tombaugh was able to locate the planet in January of 1930. The new world was christened after the God of the Underworld in keeping with the mythological tradition—and that Percival Lowell's initials might appear in the planet's name.

taken—and written up for various scientific journals by the astronomer and his associates Pickering and one Mr. A. E. Douglass—Lowell set about his dissection of the Martian surface and, in particular, its puzzling canals. First, however, let's take a look at the other Martian sights as Lowell saw them.

[Due to the tilt of its rotational axis—24°.936, an inclination within 1°.5 that of the earth—Mars has seasons which are longer than but roughly correspondent to our own. In the Northern Hemisphere, the winter averages 160 days, the spring 199, the summer 182, and the fall 146. The Southern Hemisphere enjoys a complimentary year, the summer being 160 days, the fall 199, the winter 182, and the spring 146. As one might expect, these seasons greatly affect the planet's topography. Lowell found the polar caps to be particularly susceptible to the changing weather conditions. Subject to the comparatively high temperatures of spring and summer, they melted on a regular basis. Of course, Lowell did not know the temperatures in terms of actual degrees; he simply assumed, from observation, that "the mean temperature is relatively high [for] the polar caps melt to a minimum quite beyond that of our own." The key factor was that they *did melt* and, in turn, influenced the rest of the planet.

[Lowell's attention, during this 1893–94 opposition study, was, specifically, the Southern Hemisphere. During the late spring, the striking white expanse stretched 2,035 miles across the pole, and a dark band could be seen girdling the ice cap. Lowell's powerful telescope was the first instrument ever to have perceived this feature; as deciphered by Lowell, the band—"a deep blue, like some otherworld grotto of Capri"—could have been nothing else but water at the edge of the cap, a sea created by the melting snows. However, the most important event of this opposition was that by late summer, and for the first time in the history of Martian observation, the Southern Polar Cap melted completely away. Ordinarily, it would have diminished to a few hundred miles across and be reborn again during the winter. But the surface beneath the cap and the blue band became a solid yellow expanse, and Lowell took it to be land. From this he deduced that something on Mars had to be draining the water, a suspicion he confirmed with a subsequent observation. Bordering the blue band, to the north, were grayish-green areas which Lowell at first thought were seas, but later decided were shifting patterns of vegetation—"leaves and grasses" as he described them. Some of the water was no doubt going to nourish this growth. Yet, that still did not account for the bulk of the melted ice. Nor could reddish-ochre regions that bounded these herbaceous seas have been consuming the water, since Lowell took them to be deserts: "They behave just as deserts should; behave, that is, by not behaving at all," he wrote in his book *Mars* (1894). Then, *upon* the seas, the poles, and the deserts, Lowell saw "fine lines and little gossamer filaments . . . cobwebbing the face of the Martian disk," those markings which Schiaparelli had christened canali. Perhaps, Lowell reasoned, they were ducts of some sort, channeling the water across the planet. Of course, in simply spotting them, the astronomer felt that he had done much to silence the critics of his predecessor. However, it was Lowell's photographer C.O. Lampland who ended the controversy when, in 1905, he captured thirty-eight of the canals on a single plate of film, for which he was awarded a medal by the British Royal Photographic Society.

[As to a natural or artificial origin of the canals, Lowell had this to say in *Mars*:

When the great continental areas, the reddish-ochre portions of the disk, are attentively examined in sufficiently steady air, their desert-like ground is seen to be traversed by a network of fine, straight, dark lines. The lines start from points on the coast of the blue-green regions . . . and proceed directly to what seem centres in the middle of the continent, since most surprisingly they meet there other lines that have come to the same spot with apparently a like determinate intent. And this state of things is not confined to any one part of the planet, but takes place all over the reddish-ochre regions. The lines appear either absolutely straight from one end to the other, or curved in an equally uniform manner. There is nothing haphazard in the look of any of them. Plotting upon a globe betrays them to be arcs of great circles almost invariably, even the few outstanding exceptions seeming to be but polygonal combinations of the same. Their most instantly conspicuous characteristic is this hopeless lack of happy irregularity. They are, each and all, direct to a degree.

[Upon further study, Lowell determined that the canals were an average of thirty miles wide for their entire length, which was itself a mean 1,000–1,500 miles: the longest recorded canal was Orcus, span-

46

ning 3,540 miles. And, lest anyone now or then think that Lowell was not utterly devoted to their study, he meticulously mapped and named 184 of these gossamer filaments![5] But the importance of the canals was not in their identification. As had been the case with Schiaparelli, it was in their interpretation that Lowell set the world on its ear. He very methodically ruled out natural rivers as being the cause of the canals, for their courses were *too* straight and with a common width that could not have been random. Nor could they have been cracks formed by volcanic or glacial activity, again because of this symmetry. Thus—and it was here he accounted for the vanishing water—Lowell decided that they were the work of draughtsmen—implying intelligence, as indeed he had intended. It was his hypothesis that Mars was fast dehydrating, unable to hold its seas with a gravitational force that was one third that of the earth. The canals were therefore a desperate effort by the Martian civilization to irrigate their planet with water from the polar caps. The severity of the situation Lowell inferred, in part, from the perfect straightness of the lines: not even mountains had stopped the Martians! Lowell further supported his theory by noting that the canals are visible at all because they are lined on either side with strips of vegetation. This meant that even the flora was clustering about whatever water they

5 Lowell's list of canals is as follows: Acalandrus / Acampsis / Acesines / Achana / Achates / Achelous / Acheron / Acis / Aeolus / Aesis / Aethiops / Agathodaemon / Alpheus / Ambrosia / Amenthes / Amphrysus / Amystis / Anapus / Anataeus / Anubis / Araxes / Arges / Arosis / Arsanias / Artanes / Asopus / Astaboras / Astapus / Atax / Athesis / Avernus / Avus / Axius / Axon / Bactrus / Baetis / Bathys / Bautis / Belus / Boreas / Boreosyrtis / Brontes / Caicus / Cambyses / Cantabras / Carpis / Casuentus / Catarrhactes / Cayster / Centrites / Cephissus / Cerberus / Cestrus / Chaboras / Chretes / Chrysas / Chrysorrhoas / Cinyphys / Clitumnus / Clodianus / Cophen / Coprates / Corax / Cyaneus / Cyrus / Daemon / Daix / Daradax / Dardanus / Dargamanes / Deuteronilus / Digentia / Dosaron / Drahonus / Elison / Eosphorus / Erannoboas / Erebus / Erinaeus / Erymanthus / Erynnis / Eulaeus / Eumenides / Eunostos / Euphrates / Eurymedon / Eurypus / Evenus / Fortunae / Gaesus / Galaesus / Galaxias / Ganges / Ganymede / Garrhuenus / Gigas / Gihon / Glaucus / Gorgon / Gyes / Hades / Halys / Harpasus / Hebe / Helisson / Heratemis / Herculis Columnae / Hiddekel / Hipparis / Hippus / Hyctanis / Hydaspes / Hydraotes / Hydriacus / Hylias / Hyllus / Hyphasis / Hypsas / Hyscus / Indus / Iris / Isis / Jamuna / Jaxartes / Labotas / Laestrygon / Leontes / Lethes / Liris / Maeander / Magon / Malva / Margus / Medus / Medusa / Mogrus / Nectar / Neda / Nepenthes / Nereides / Nestus / Neudrus / Nilokeras / Nilosyrtis / Nilus / Nymphaeus / Oceanus / Ochus / Opharus / Orcus / Orontes / Orosines / Oxus / Pactolus / Padargus / Palamnus / Parcae / Peneus / Phasis / Phison / Protonilus / Psychrus / Pyriphlegethon / Rha / Scamander / Sesamus / Simois / Siernius / Sitacus / Steropes / Styx / Surius / Tartarus / Tedanius / Thermodon / Thyanis / Titan / Tithonius / Triton / Tyndis / Typhon / Ulysses / Uranius / Xanthus /

could find. As he so inimitably expressed it in *Mars*, "Across the gulf of space that separates us from Mars, an area thirty miles wide would be perceptible as a dot. [But] if there be a scarcity of water upon the surface of the planet, the necessary water would have to be conducted to the mouths of the canals across the more permanent areas of vegetation, thus causing bands of denser verdure athwart them." In addition to the regular canals, Lowell saw what he called *double canals*, channels parallel to the more prominent lanes, with approximately 150 miles between them. These were transient features and visible only under the most exemplary conditions: Lowell had no justification for their appearance other than to be "tolerably sure that the phenomenon is not only season but vegetal," and admit that an explanation would not be easily rendered.

[However, in Lowell's judgment, his analysis of the origins of the canals was clinched by the way that both the single and double lines all ran to 120–150 mile-wide junctions or oases in various sections of the planet. Since the oases presented a convex rather than concave face to the entering canals, they could not have been simply expansions of the canals. Thus, he interpreted them as having first been pockets of water and, when they went dry, serving as transfer stations for water from the poles. In all, Lowell and his associates identified sixty-three oases.[6]

[What did these various discoveries mean to Lowell? Well, in *Mars*, he was already referring to Martians as old friends.

A mind of no mean order would seem to have presided over the system we see—a mind certainly of considerably more comprehensiveness than that which presides over the various departments of our own public works. Party politics, at all events, have had no part in them; for the system is planet-wide. Quite

6 Lowell's list of oases is as follows: Acherusia Palus / Aganippe Fons / Alcyonia / Ammonium / Aponi Fons / Aquae Apollinares / Aquae Calidae / Arachoti Fons / Arduenna / Arethusa Fons / Arsia Silva / Arsine / Astrae Lacus / Augila / Bandusiae Fons / Benacus Lacus / Biblis Fons / Castalia Fons / Ceraunius / Clepsydra Fons / Cyane Fons / Cynia Lacus / Daphne / Ferentinae Lucus / Flevo Lacus / Fons Juventae / Gallinaria Silva / Hercynia Silva / Hesperidum Lacus / Hibe / Hippocrene Fons / Hipponitis Palus / Hypelaeus / Labeatis Lacus / Lacus Ismenius / Lacus Lunae / Lacus Phoenicis / Lausonius Lacus / Lerne / Lucrinus Lacus / Lucus Angitiae / Lucus Feronia / Lucus Maricae / Maeisia Silva / Mapharitis / Mareotis / Meroe / Messeis Fons / Nitriae / Nodus Gordii / Nessonis Lacus / Oxia Palus / Palicorum Lacus / Pallas Lacus / Propontis / Serpium / Sirbonis Lacus / Solis Fons / Solis Lacus / Tithonius Lacus / Trinythios / Trivium Charontis / Utopia /

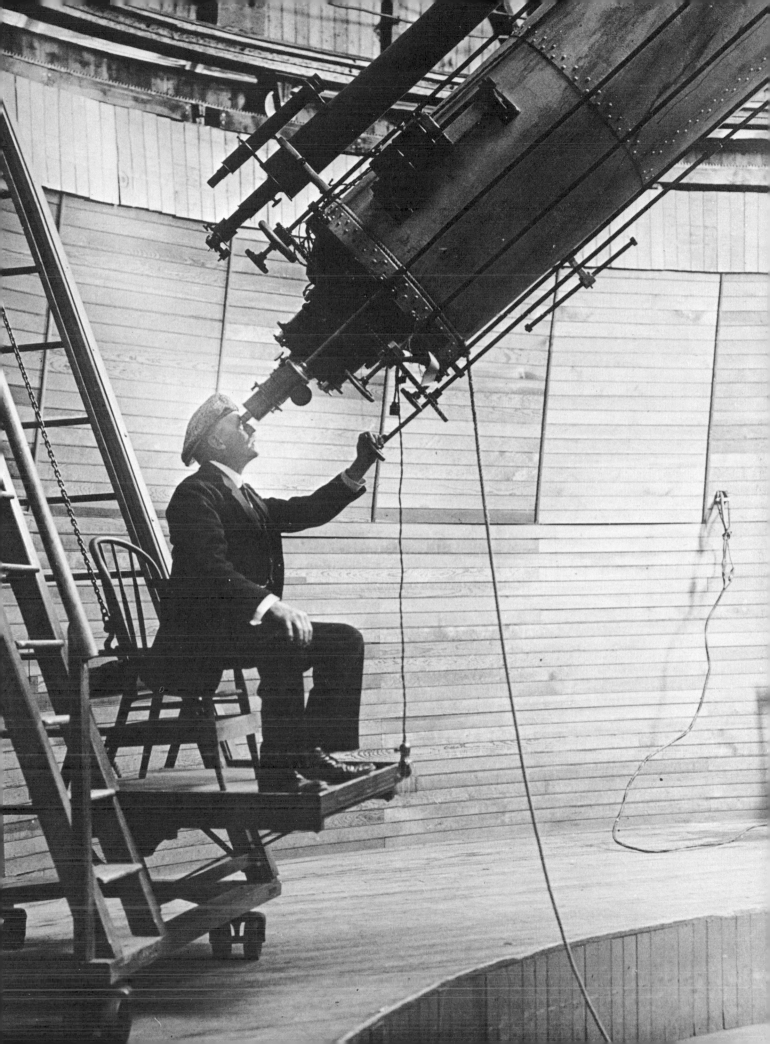

possibly, such Martian folk are possessed of inventions of which we have not dreamed, and with them electrophones and kinetoscopes are things of a bygone past, preserved with veneration in museums as relics of the clumsy contrivances of the simple childhood of the race. Certainly what we see hints at the existence of beings who are in advance of, not behind us, in the journey of life.

However, the astronomer was cautious to add,

To talk of Martian beings is not to mean Martian men. Just as the probabilities point to the one, so do they point away from the other. Even on this earth, man is of the nature of an accident. He is the survival of by no means the highest physical organism. He is not even a high form of mammal. Mind has been his making. For aught we can see, some lizard or batrachian might just as well have popped into his place early in the race, and been now the dominant creature of this earth. Under different physical conditions, he would have been certain to do so. Amid the surroundings that exist on Mars, surroundings so different from our own, we may be practically sure other organisms have been evolved of which we have no cognizance. What manner of beings they may be we lack the data even to conceive. For answers to such problems, we must look to the future.

[But when the writers of Lowell's day fused his theories with the stigma of the war gods, it appeared that our earth should *have* no future! Not only did the Martians infiltrate or raze our world, but they plucked away the flower of our manhood to partake in deadly adventures on the Red Planet. We'll be returning to Lowell in chapter four. Right now, however, let's look at some drama and a bit of fluff. . . .
[Not surprisingly, the work of Schiaparelli and Lowell made Mars a bankable commodity. Reporters, lecturers, and especially scientists were able to commercialize on the planet either by reviewing, discussing, or debunking the astronomers' findings. Conversely, the foremost concern of the earliest Mars-probing novelists was to come up with viable entertainment. And while these initial efforts may appear crude, remember that they were, after all, trailblazers. Percy Greg's *Across the Zodiac* (1880) was the first such book, the droll

account of a planet-hopping engineer. To the best of anyone's knowledge, his spaceship *the Astronaut* was well suited for the voyage: twenty feet tall and one hundred feet long, it was powered by a propulsive force known as apergy,[7] and boasted metal walls over three feet thick. Sailing into space, the craft transports our flyer to Mars where he finds a civilized race of beings. The visitor spends a pleasant time on their world, but must leave after offending planetary sovereignty by becoming too deeply embroiled in Martian matters.[8] Eleven years later, author Robert Cromie told of humankind's second assault on the Red Planet via the spherical vessel *Steel Globe*, launched from a desolate Alaskan tundra. Capable of controlling or nullifying the force of gravity, the spaceship carries a handful of men to Mars. There, they discover a sophisticated society whose members lead a life of ease and unconcern. Indeed, with the exception of a love affair between one of the astronauts and a Martian miss, nothing much happens until the men return to earth. On the last leg of their journey, they discover that their oxygen is being consumed far too quickly. The cause? Their comrade's alien girlfriend has snuck onboard to be with her lover. Realizing that her impetuous move has endangered the crew, the girl willingly sacrifices her life, stepping through an airlock into space.

[Clearly, these seed works are important as historical relics, but they are hopelessly superficial as studies of the Mars-humankind relationship. To provide insight or drama, one needs *conflict*—and, as though apologizing for its predecessors, the next Mars book had an abundance of mayhem! Crafted with supreme style it remains, to this day, a classic. Herbert George Wells (1866–1946) was thirty-two when he penned *War of the Worlds*, the first and still the finest novel about an attack from the Red Planet. Wells' Mars is similar to the planet that Lowell described; an old world, wise but dying. In fact, the words used by Wells to begin his tale might just as readily have come from Lowell:

No one would have believed in the last years of the nineteenth century that this world was being watched keenly and closely by intelligences greater than man's and yet as mortal as his own;

7   It's conceivable that "apergy" is a play on the word "apogee," the point in a body's orbit when it is farthest from the earth.
8   A sequel, published in 1894, sent our spaceman to Jupiter and Saturn onboard the apergy-powered *Callisto*.

that as men busied themselves about their various concerns they were scrutinised and studied, perhaps almost as narrowly as a man with a microscope might scrutinise the transient creatures that swarm and multiply in a drop of water. With infinite complacency, men went to and fro over this globe about their little affairs, serene in their assurance of their empire over matter. It is possible that the infusoria under the microscope do the same. No one gave a thought to the older worlds of space as sources of human danger, or thought of them only to dismiss the idea of life upon them as impossible or improbable.

[Wells continues his narrative, telling us that flashes seen upon the surface of Mars were actually the launching of artificial projectiles toward the earth. Days later, one of these cylinders, thirty yards wide and hollow, impacts on a green just outside the town of Woking, England. Several hours later, humankind meets its first Martian. Wells' nameless hero and Narrator, a writer of philosophy, describes the event thusly:

I looked again at the cylinder and ungovernable terror gripped me. I stood petrified and staring. A big grayish rounded bulk, the size, perhaps, of a bear, was rising slowly and painfully out the cylinder. As it bulged up and caught the light, it glistened like wet leather. Two large dark-coloured eyes were regarding me steadfastly. The mass that framed them, the head of the thing, it was rounded, and had, one might say, a face. There was a mouth under the eyes, the lipless brim of which quivered and panted, and dropped saliva. The whole creature heaved and pulsated convulsively. A lank tentacular appendage gripped the edge of the cylinder, another swayed in the air.

Those who have never seen a living Martian can scarcely imagine the strange horror of its appearance. The peculiar V-shaped mouth with its pointed upper lip, the absence of brow ridges, the absence of a chin beneath the wedge-like lower lip, the incessant quivering of this mouth, the Gorgon groups of tentacles, the tumultuous breathing of the lungs in a strange atmosphere, the evident heaviness and painfulness of movement due to the greater gravitational energy of the earth—above all, the extraordinary intensity of the immense eyes—were at once vital, intense, inhuman, crippled and monstrous. There was something fungoid in the oily brown skin, something in the clumsy deliberation of the tedious movements unspeakably nasty. Even at this first encounter, this first glimpse, I was overcome with disgust and dread.

And with good reason! No sooner do the Martians emerge than they turn a vaporizing ray on those who had gathered about the cylinder. The Narrator survives this attack and tells of the chaos which ensues. More Martians arrive in cylinders and overcome the tug of our gravity by mounting their invasion from within huge fighting-machines. Each one of these "monstrous tripods [was] higher than many houses . . . a walking engine of glittering metal [with] articulate ropes of steel dangling from it." Equipped with the awesome incendiary ray, the robots destroy forests, buildings, and even ships with easy nonchalance. Human beings, if they are not cremated, are tossed into a great metallic carrier—"like a gigantic fisherman's basket"—lashed to the back of each Titan, preserved so that their blood can be drained to feed the Martians. City after city falls, and with the crumbling of the armed forces and terrestrial order, it appears to be all over for humankind. The Martians would soon be re-established on a planet rich in seas and sustenance. Then, as the Narrator resolves to die before one of the three-legged giants, rather than be bred as food, the tripods go still. And "in their overturned war-machines . . . were the Martians—*dead!*—slain by the putrefactive and disease bacteria against which their systems were unprepared . . . slain, after all man's devices had failed, by the humblest things that God, in his wisdom, had put upon the earth."

[Before we examine Wells' novel as the record of humankind's first contact with Martians, it should be noted that *War of the Worlds* is not so much a work of plot and subplot as it is a brilliant study of theme and simile. We, ourselves, are the Martians; in particular, Wells' metaphoric target was Great Britain and its vicious imperialism in Africa, India, Asia, and portions of the Far East. For this reason, the author constantly refers to the Martian-human relationship as being equitable with a human-ant, human-beast, or human-monkey comparison. In fact, in describing the Martians' taste for human blood, the Narrator wonders "how repulsive our carnivorous habits would seem to an intelligent rabbit." It is a rhetorical query which puts the invasion very neatly into perspective. Wells furthers his analogy by noting that humankind has wiped out animals like the bison and the dodo as well as "inferior" races such as the Tasmanians; he therefore concludes, "Are we such apostles of mercy to complain if the Martians warred in the same spirit?" However, this attempt to put a haughty race in its place returns us to the subject itself. That the novel succeeds is due to Wells' utterly realistic portrait of the rampaging Martians and the futility of the human lot. The characters are brilliantly sketched, their panic and fear effectively described as the result of trying to fight lightning with arrows. Too, Wells is careful to root all speculation in data that is consistent with what was known or surmised about Mars. He makes numerous references to the work of Schiaparelli and the famous observatories of his day, and takes pains to explain Martian logic and science wherever possible. Indeed, Wells' thoroughness helps to show the Martians as those creatures which we ourselves sometimes become, soulless beasts adroitly personified by the following description of the alien visceral system:

> The greater part of the structure was the brain, sending enormous nerves to the eyes, ear, and tactile tentacles. Besides this were the bulky lungs, into which the mouth opened, and the heart and its vessels . . . the complex apparatus of digestion, which makes up the bulk of our bodies, did not exist in the Martians. They were heads—merely heads. Entrails they had none. The physiological advantages . . . are undeniable. Men go happy or miserable as they have healthy or unhealthy livers, or sound gastric glands. But the Martians were lifted above all these organic fluctuations of mood and emotion.
>
> In three other points their physiology differed strangely from ours. Their organisms did not sleep, any more than the heart of a man sleeps. Since they had no extensive muscular mechanism to recuperate, that periodical extinction was unknown to them. In the next place . . . the Martians were absolutely without sex, and therefore without any of the tumultuous emotions that arise from that difference among men. A young Martian [was] born upon earth during the war, and it was found attached to its parent, partially *budded* off, just as young lily-bulbs bud off, or like the young animals in the freshwater polyp. The last salient point . . . one might have thought a very trivial particular. Micro-

54

organisms, which cause so much disease and pain on earth, have either never appeared upon Mars or Martian sanitary science eliminated them ages ago.

Ironically, of course, it is this sophistication which ultimately destroys the Martians.
[The clever symbolism aside, Wells also contributes other Martian life-forms to our literary search for natives of the Red Planet. He mentions one type of fauna as a cogent accent to his slightly cynical motif:

[The Martians'] undeniable preference for men as their source of nourishment is partly explained by the nature of the remains of the victims they had brought with them as provisions from Mars. These creatures, to judge from the shriveled remains that have fallen into human hands, were bipeds with flimsy, silicous skeletons (almost like those of the silicous sponges) and feeble musculature, standing about six feet high and having round, erect heads, and large eyes in flinty sockets.

. . . and one type of flora.

Apparently, the vegetable kingdom in Mars, instead of having green for a dominant colour, is of a vivid blood-red tint. At any rate, the seeds which the Martians . . . brought with them gave rise in all cases to red-coloured growths.

It is indeed appropriate that the natural color of Mars so readily lent itself to the many blood references used by Wells!
[Because of this depth, articulate telling, and vivid imagery, *War of the Worlds* is the Martian vintage by which all others are judged. It was particularly striking because, in the late nineteenth and early twentieth century, the story was unprecedented, and the market devoid of competition. Too many Martians would have taxed the interest of the book-buying public, and digest-size science fiction magazines, geared to fans of the genre, did not begin publication until 1926. Thus, the next Red Planet epic, *The Pharaoh's Broker* (1899), is of an entirely different timbre than Wells, Greg, or Cromie. It's a satire in which Jewish grain merchant Isidor Werner and physicist Hermann Ander-

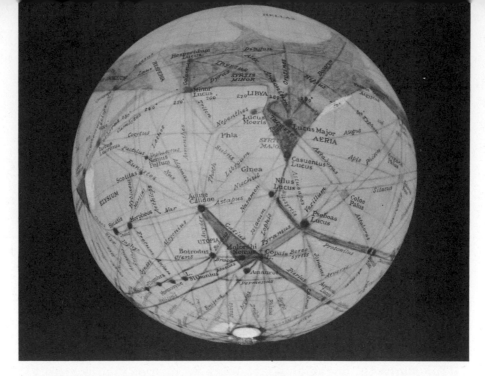

The globe of Mars built by Percival Lowell, showing
many of its canals (*A Lowell Observatory photograph*).

welt take a trip to Mars. Voyaging through space in a rocketship
financed by Werner, the men learn that the civilization on Mars
closely parallels that of Ancient Egypt. And if this is so, the mercenary
Werner's reasons, then history will repeat itself and the country is
due for seven years of famine. Hoping to make a fortune during this
period, Werner monopolizes the planet's wheat market, incurs the
Martians' dander, and both he and Anderwelt are forced to depart.
The terrestrials can't even use the wealth they have accumulated
since the currency on Mars is iron! However, from their expedition, the
men deduce that each planet goes through a historical evolution
similar to that of the world before it. Therefore, a trip to Venus is
planned to learn about the earth's future; unfortunately, author
Ellsworth Douglass never wrote the tale.
[Moving from the serio-comic and past some bland Martian fan-
tasies, we find the paradox of *Lieutenant Gulliver Jones: His Vacation*
by Edwin Lester Arnold. Published in 1905, this obscure and unimaginative little novel merits our attention because it served as the
conceptual blueprint for the most successful Martian series of all time.
First, however, let's look at Lt. Jones. The term "romantic" is applied
to science fiction when credible technology is ignored in favor of bold,
heroic idealism and exotic adventure. Thus, as opposed to the sly
chronicle of Douglass, the parable of Wells, and the technical novels
of Greg and Cromie—with their interesting emphasis on the hardware

used to travel to Mars—Arnold's work is the first Martian fairy tale. Despondent over his inability to win the lovely Polly, Lt. Jones, on leave from the Navy and standing atop a rug newly laid in his New York apartment, wishes he were anywhere but on this frustrating world. He absently suggests that the planet Mars might be a proper place to visit. Moments later, what turns out to have been a *magic* carpet rolls the sailor within its folds, kicks about the apartment, then darts out the window toward Mars. The lieutenant passes out. When he awakens, Gulliver finds himself surrounded by black-and-red hills, a thin opal-colored mist, and dawning sky of gold and red. Welcomed to Mars by ivory-colored, perpetually even-tempered humanoids, Gulliver is led to the Martian city of Seth, a once-glorious township that has fallen on hard times. The natives teach Gulliver their lisping tongue by osmosis, and when he saves the lovely Princess Heru from drowning, Gulliver is invited to stay at the court of her suitor, Prince Hath. Almost immediately, however, the visitor creates a problem as Heru rigs the traditional mass marriage, in which mates are selected by chance, so that she is paired with Gulliver. This irritates Hath, but there is no time for him to react: after the ceremony, the court is visited by three ape-like men, short, broad shouldered, and with copper-colored skin. They have come from the woodlands across the sea to collect an annual tribute for their powerful leader Ar-hap. And, when Gulliver falls asleep, the bullet-headed emissaries also collect Heru. Gulliver learns of the kidnapping and sets out after his princess. The journey is fraught with the peril of wolves, bears, a huge Martian elk, giant bats, gulls with forked tails, tiny lizards with wings of transparent green, and elephant-size rats; it is also dressed with the strange beauty of alien plants, flowers which change color from hour to hour, vines which grow so quickly that to sit near them was to become hopelessly entangled, and buds that cry out with the voice of a sobbing maiden. Eventually, Gulliver makes his way through forests, across tumultous seas, and along a river sided by cliffs of blue ice in which eons of Martian dead are enclosed—"a huge amphitheatre of fossilized humanity," as Gulliver describes it—to the western lands of Heru's abductors. Arriving onboard a merchant's ship, Gulliver's presence becomes known to Ar-hap, who summons the newcomer to an audience. Gulliver demands Heru's release and the ruler agrees, but only if the alien will perform two impossible deeds. The king names them; Gulliver works the wonders and Heru is his.

The lovers board a canoe and venture downstream. However, Ar-hap had never intended to surrender his prize. He and his army pursue the pair to Seth and surround the palace. Although Gulliver is able to sneak Heru and a handful of slaves onto a boat bound for another settlement, he does so at great cost to Seth: the angry Ar-hap puts the city to torch, Hath is murdered, and Gulliver, barricaded in a storage room, is besieged by 50 of the gorrilla-men. They break down the wooden door and the earthman, looking for something to throw at them, spots the magic carpet. Throwing himself upon its back, Gulliver wishes that he were back in New York and is spirited away. The lieutenant is deposited before his apartment and finds, in the mail which arrived during his absence, a notice of promotion. His income and station greatly improved—and his love for the beautiful Heru evidently forgotten—Gulliver asks Polly to marry him. She consents, asking but a single favor: "You shall write a book about that extraordinary story you just told me." And so it is that we come to have Gulliver's tale.

[The fact that Gulliver Jones lived, loved, and fought on Mars seems to have made little impression on the author. Thus, while one can be caught up in the spirit and innocence—some might call it naivete—of the novel, one must also lament its sorry lack of substance or resourcefulness. Nowhere is there any mention of Mars' lesser gravity or alleged canals; nothing is made of Gulliver's observations about the planet's thin atmosphere. In short, the setting was dictated by Mars being so much in the news, and not because it inspired or held any scientific allure for the author. Gulliver's story could just as easily have taken place in a Norse village or the American West! Happily, though, the novel is kept buoyant by the fact that there is none of the selfless hero in Gulliver. En route to rescue Heru, for example, he wonders—more than once—why "I am here on a hare-brained errand, playing knight-errant in a way that shocks my common sense?" It is an added dimension that coaxes the reader through many prosaic stretches and an abominable lack of generic Martian sights, sounds, and character. However, if Gulliver Jones is a reluctant hero, and his Mars is a mere shadow of what it could have been, then Edgar Rice Burroughs' John Carter is his antithesis, the ultimate hero on a world of which Percival Lowell would have enthusiastically approved.

[Edgar Rice Burroughs (1875–1950) failed at numerous occupations before writing his first novel in 1911. Entitled *Under the Moons of*

Mars, it was published the following year in the *All-Story* pulp magazine. This was the periodical which, later that year, would also run what was to be Burroughs' most popular novel, *Tarzan of the Apes. Under the Moons of Mars*—henceforth referred to as *A Princess of Mars*, the title of its first and subsequent book publications—is the muscular saga of Capt. John Carter who, facing certain death at the hands of Apache warriors, is astral transported to the planet Mars. The world that welcomes Carter is remarkable in its beauty, danger, and mystery. Ruled at one time by the albino Orovars, whose great fleets roamed the five Martian seas and kept peace throughout the planet, Mars is now a decaying planet. A million years before Carter's arrival, the oceans began to evaporate and the power of the Orovars faded. Subjugated races rebelled; particularly ruthless were the 15-foot-tall *tharks*, hairless green barbarians sporting four arms and a pair of upturned tusks. As a result of these uprisings, the Orovar civilization was destroyed. Most of those who managed to escape were assimilated into the races of orange-skinned and black-skinned humanoids, thus breeding the copper-red nation which wields the balance of power at the time of Carter's arrival. *Princess of Mars* wastes little time giving the "trained fighting man" a challenge on which to whet his skills. No sooner has he become accustomed to the lesser gravity on Mars, which gives him superhuman strength and allows him to leap 10 feet into the air, than Carter is attacked and taken prisoner by a band of tharks astride *thoats*, swift, eight-legged mounts 10 feet tall at the shoulder that are reined by telepathic thoughts from the rider. Carter is taken to the thark community where his assistance in repelling a herd of white, four-armed apes earns him the respect of his captors. They teach Carter the language and instruct him in the use of bizarre Martian weapons. In the meantime, another human being is captured, the incomparable Dejah Thoris, the titular princess of Mars—or, as the natives call their world, *Barsoom*.[9] Carter falls in love with the girl, and together they flee the thark city, becoming involved in various adventures. Burroughs, however, chose to leave his readers clamoring for a second installment: after marrying Dejah and fathering an as yet unhatched egg—all Martians are egg-laying creatures, according to the author—the atmosphere plant

9 No one knows just how Burroughs came up with this name. Although students of the author might dispute this analysis, I see "Barsoom" as having possibly come from Burroughs' tinkering with the words "Mars" and "moon."

which restocks the attenuated Martian air blanket breaks down. Carter boards a flier—an antigravity transport which can travel two hundred miles per hour, a legacy of Orovar science—to help workmen try and repair the station. But the prince is mysteriously drawn back to earth before he can accomplish his goal. Thus, not only has he been taken from Dejah Thoris, but he does not even know if she is still alive! This, quite clearly, called for a sequel, which *All-Story* showcased in 1913. In *The Gods of Mars*, Carter returns to Barsoom and is set upon, not by tharks, but by a *banth*: a huge, 10-legged carnivore roughly equivalent to the terrestrial lion. Clearly, Carter's fellows had succeeded in repairing the pumps! However, Dejah Thoris is again in trouble, kidnapped by the black pirates. With the aide of his grown son, Carthoris, Carter ventures down the Iss, a river along which dying Martians make their final pilgrimage, to the South Pole. There, Dejah is a prisoner of the flesh-eating Therns, descendents of the Orovars and keepers of the faith of Issus, Daughter of the Lesser Moon. Here too, incidentally, are more pumping stations, moving channel water from an underground sea to the ice cap which, when it melts, supplies the rest of Barsoom with water via canals. Once again, Burroughs leaves his readers with a cliffhanger: with a stone wall between them, Carter is unable to save Dejah from the knife of Phaidor, a rival for his affections. However, in Book Three, *The Warlord of Mars* (1914), we learn that Dejah Thoris did not die. Thuvia of Ptarth, a slave of the Therns and another of John Carter's female admirers, intercepts the blade of the daughter of the High Priest of Issus, thus preventing the murder. However, to free his beloved, our hero must follow her trail to the North Pole, across crimson plains and the Valley Dor, through the land of the carnivorous *calot-trees*, the *apt*—a four-legged, two-armed, white-furred hippopotamus—the deadly *sith*—bull-sized hornets—to Okar, home of the yellow men and their merciless Jeddak (prince) Salensus Oll. With the aide of Carthoris, Tars Tarkas—a Thark who had been at John Carter's side since Book One—and Woola—Carter's pet *calot*, a pony-sized, six-legged dog—Carter rescues his wife from the Jeddak who would be her mate, and is named Warlord of Barsoom, the highest of all Martian honors. And so ends the John Carter trilogy. But the public wanted more stories of the Red Planet. Thus, besides writing 25 Tarzan adventures, the "inner earth" Pellucidar series, the Carson of Venus books, and some 20 other novels, Burroughs recounted the adventures of Carth-

60

oris for a fourth book, *Thuvia, Maid of Mars* (1920); the escapades of John Carter and Dejah Thoris' daughter Tara in *Chessmen of Mars* (1922); the battles of John Carter and his terrestrial friend Ulysses Paxton as they work to destroy Ras Thavas, the insidious *Mastermind of Mars* (1928); the story of Hadron of Hastor, a Carter aide, in *A Fighting Man of Mars* (1931); John Carter's bout with an assassination guild in *Swords of Mars* (1936); a second encounter with Ras Thavas in *Synthetic Men of Mars* (1940); the struggles of the Warlord against a race of subterranean Orovars in *Llana of Gathol* (1948)[10]; and finally, in the posthumously published *John Carter of Mars* (1963)—the collective title for two novelettes—our hero's fight with Pew Mogel and his monstrous leviathan Joog in *John Carter and the Giant of Mars* (1940), and with beings from another world in the *Skeleton Men of Jupiter* (1943).

[Beyond the marvelous complexity of these novels—the entire Martian tapestry runs nearly 700,000 words—and the incredible names, games, rituals, and language—all of which Burroughs explains in painstaking detail—the joy of Barsoom[11] lies in its devotion to the Mars of Schiaparelli and Lowell. Not only does the author work the canals, the thin atmosphere, the rusty color, the polar caps, and the low gravity of Mars into his narratives, but he illustrates how the history, flora, and fauna of the planet were influenced by these environmental elements. However, more important than the author's diligence and his impeccable sense of plotting, is that the series pays tribute to the noble Mars of both scientific speculation and the dreams of Burroughs' readers. It is respectful—not condescending. And, while ideas from the Edwin Arnold novel may have worked their way into the John Carter books—such as the River of the Dead, the mystical route to Mars, the motif of nomadic tribes and their hatred for city-dwellers, and even John Carter himself, cast from the mold of Arnold's *Phra the Phoenician* (1890)—it must be understood that Burroughs came to writing with no training or experience whatsoever. He was uncertain, impressionable, and deserves great credit for even realiz-

10 Llana of Gathol is the daughter of Tara and, hence, John Carter's granddaughter.
11 Although "Barsoom" was a word easily enough dissected, it would take a truly devoted Burroughs scholar to search for the roots of the author's every Martian term. One fan dictionary lists nearly 150 words, from Aaanthor (a dead city of ancient Mars) to Zode (a Martian hour). This is just one example of the detail and order in Burroughs' universe. And consider that he had a separate and distinct language for each of his literary series!

61

ing the potential of these tantilizing concepts.[12] In a larger, more important sense, he was the man who may very well have put us on the moon! More on this thought in chapter four.

[Despite the popularity of Burroughs, there was more than swashbuckling to be done on Mars in these early years of the twentieth century. In 1911, Hugo Gernsback began running fiction pieces in his magazines *Modern Electrics* and *Science and Invention*. As one can well imagine, the mechanical heroes (and writing!) of these tales generated more voltage than literature. But the stories were enthusiastically received, and inspired Gernsback to found *Amazing Stories* in 1926. Inevitably, Gernsback offered his readers the same gadget-laden fiction that had appeared in the technical journals. However, he abetted his early and unrounded "hardcore" fiction with reprints of Verne, Wells, Burroughs, and others. The result was a balanced and intriguing package. Sales increased steadily and, before long, *Amazing* was able to eliminate most of the reprint material. Gernsback expanded the scope of his stories to include characterization and plot and, during the twenties, thirties, and forties gave us dozens of tales about Mars. And, although study of the planet had barely progressed beyond the discoveries of Percival Lowell (see chapter four), the *Amazing* selections did not wallow in previously explored themes. In 1926, Robert H. Goddard had launched the first liquid-fuel rocket, a development which gave subsequent outer-space fiction much of its wonder and credibility. Here, at last, was a *sensible* way to get to the moon or Mars, and it was a tool widely embraced by science fiction authors. Needless to say, there were also the everpopular Martians on earth: lots of 'em, since they made for interesting, often humorous storytelling. In short, these magazines— *Amazing* was joined by such Gernsback titles as *Science Wonder Stories* and *Air Wonder Stories*, and the rival publications *Astounding* and *Startling Stories*—were a showcase that not only *tolerated* the fantastic: they were *devoted* to it! Let us look, then, at some of the more representative Martian tales.

[By 1990, says Thornton Ayre, author of *Locked City*, the evil dictator Baxter Holroyd will rule the earth. And the punishment he orders for captured revolutionaries Rod Calab, his wife Eva, and Boris

12   Today, the Burroughs books and merchandising is handled by the author's estate in Tarzana, California. However, there has yet to be a motion picture or product that does justice to the work of Edgar Rice Burroughs.

Rengard, is to fly the first manned rocket into space. Their destination, of course, is Mars. Arriving on the Red Planet after a two-week flight (!), they find a magnificent city covering several square miles, surrounded by forests and fed by canals. Within the metropolis proper are monorails, buildings of sleek, blue metal—and seven blue-skinned children. These youngsters are the sole inhabitants of Mars, and the earthlings learn that they were test tube babies, created by the Martians before their fresh drinking water ran out. The youths survive by eating only moist fruits which filter the impurities from nearby lakes—lakes consisting of a deadly water isotope. Besides preserving the race, however, the children have been charged with maintaining the fully automated city for their own future. Rod sees this keepership as a godsend: using their knowledge, he makes the Martians' robot slaves and rocket fleets operational and, returning to earth, overthrows the sadistic Holroyd regime. Justice has been served! There's much more to the story, of course: detailed explanations of Martian machinery, a study of the thin atmosphere, a chemical analysis of the toxic water (complete with footnotes describing the appropriate elemental processes), and a rather sterile tour of the alien city. In other words, it's a typical Gernsback story, extending our knowledge of Mars to fictional but not implausible limits. (It's also a retelling of the Biblical Exodus, if one is inclined to make a proverbial mountain out of a mole hill. The liberator *was* banished by a cruel dictator, and he *did* return to conquer with the heavenly hosts at his side. And, of course, one of Moses' aides was named *Caleb*. . . .) Another vivid tale of life on Mars was Stanley G. Weinbaum's *A Martian Odyssey*, one of the most popular science fiction stories of all-time. Exhibiting a structural approach that is different from that of *Locked City*, *A Martian Odyssey* is the recollection of one man who was part of the first expedition to Mars. After a successful landing on the Red Planet, our hero strikes off on his own and, encountering a variety of Martian creatures, befriends one of them, the amiable, road-runnerish Tweel. Though the two beings cannot converse, they communicate through action and expression. This permits Weinbaum to crystallize the magic of eventual contact between a man and an intelligent alien. Obviously, then, *A Martian Odyssey* is the antithesis of *Locked City*: the insipid trio and scientific detail of Ayre's tale—appropriate though they were—have been replaced with people of flesh and blood and only peripheral technology, a factor which has kept the Weinbaum work alive. Machines can become outdated; people never go out of style. And, while science fiction writers would learn to draw from both these orientations (see chapter four), it was a development which lay beyond the careers of either Weinbaum or Ayre. Then again, not every pulp science fiction story could progress to even these primitive levels of accomplishment. The uninspired *Empress of Mars* is one such example. Set on a war-torn, barbarian-ridden, exotically-accoutered, canal-gridded, and otherwise redundant Mars, Ross Rocklynne's work is an unimaginative rehashing of Edgar Rice Burroughs, complete with a sword-swinging hero in Darak of Werg, the incredible beasts—the *Yammir*, a fast-flying nocturnal bird; the *Wachin*, a tiny, transparent animal with violet veins and luminous red hair; the *Hoepx*, a horned beast which uses the protrusion like a sword; etc.—and a stock plot.

[One of the more offbeat humans-on-Mars stories was *Vengeance from the Void* by Frederic Arnold Kummer, Jr. Apart from the expected melodrama, Kummer's story is unique in that it presents a history of the Martian deserts as well as the canals. Erik Steinson, pushed from a rocketship by Jarth, his uncle and copartner in the Martian Reclamation Company, is found floating in space some 15 years later. Perfectly preserved, Erik is brought back to life by Dr. Marcus Thain. Jarth, meanwhile, has set himself up as a ruthless overlord on Mars. His ace in the hole: water. Millennia before the coming of the terrestrials, the Martians had allowed their iron buildings and machines to rust, an oxidizing process which created the planet's great red deserts and transformed Mars' water into gaseous hydrogen. The canals, of course, were a last-"ditch" effort to procure water from the poles, an action which came too late to save the Martians. When humans landed on Mars, they immediately formed the MRC and began bonding hydrogen and oxygen to create water. Jarth, however, used his ownership of the firm to become the wealthy master of Mars, rationing water at great expense to the colonists. Naturally, the revived Erik topples the fiend from this heinous height of power, and all ends happily. As ever, the chemistry and biology are all dutifully documented as *Amazing Stories* does its scientific damnedest to please!

[If these outer-space tales cover a wide spectrum in both quality and originality, the difference between the *Amazing* Martians-on-earth tales and the trendsetting work of H. G. Wells is equally great!

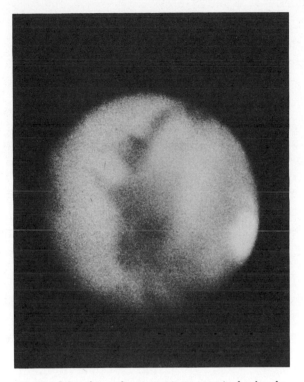

Above: Like the pulp magazines, comic books devoted a great deal of attention to Mars. This sample panel is from *Weird Science*, a publication of the early fifties (*Copyright by William M. Gaines*). *Below:* Mars as Lowell would have seen it through his telescope.

SOON, THE FLAME-RED SURFACE OF THE FOURTH PLANET FROM THE SUN IS RUSHING UP AT US...

Sometimes the aliens are bug-eyed monsters descended from *War of the Worlds*; on other occasions they're Homo sapiens. For instance, Mark Shean, of *Visitor to Earth* by P.F. Costello, is a humanoid. But he is wont to pronounce his name without the 'k' and, after spending several days inspecting our world, drops all pretense. He is really Ang-Ar of Mars, and he has a super weapon to give to the United States. Unfortunately, the weapon—a disintegration beam—falls into enemy hands, and Ang-Ar must get it back. When this is accomplished, he realizes that, although we're the "good guys," it's foolish to leave so potent an armament on earth. However, before leaving for Mars Ang-Ar remarks, "Should you really need us, we will be ready and waiting." Other Martians were not so beneficent. When a cancerlike plague strikes the Red Planet in Miles J. Breuer's *The Raid from Mars*, its thin, humanoid inhabitants search for a cure. They spy on the earth and decide that radium, the treatment used by terrestrial doctors, seems ideal. Since the Martians live underground and know from neither radiation or this particular element, they dispatch a warship to steal our supplies. However, like the *War of the Worlds* Martians, these creatures make a fatal mistake: the radium that heals can also destroy. Recklessly snatching the substance from hospitals, they are killed by the attendant radiation. Of course, not all Martians come to earth by such conventional means. Some are spirited from their native world, as is the dormant life cell found buried beneath the sands of Mars in Polton Cross' *The Martian Avenger*. Although the specimen has every appearance of being a plant seed, Dr. Lance Halworthy decides to fertilize it with human sperm cells. The egg becomes a snake-like creature, and, having a hereditary memory of the Mars of its parents, telepathically replays these visions for Lance. Apparently, a scientific race had dominated the earth long before the advent of recorded history. However, when a terrible upheaval robbed the planet of its water, our ancestors had no choice but to drain the Martian seas, which caused the famous Biblical Flood. Meanwhile, the Martians were now drought-stricken and (naturally!) built canals in order to survive. But the population was quickly reduced to a handful of scientists who, before their death, preserved the cell which Lance discovered, hoping that it would one day be nurtured and enabled to wreak vengeance against the earth. Accordingly, the alien duplicates Lance's body, shackles the original in a cellar, and goes about its mission of poisoning the earth's atmosphere. Fortunately, Lance's girlfriend becomes aware of the substitution and frees her lover who, in turn, slays the Martian. However, some Martians don't take their grudges as much to heart as *The Martian Avenger*. In *Strictly Formal*, by Gerald Vance, all is forgiven as social outcast Sam Evans receives a message from Mars, sent via meteor. Unfortunately, he spends too much time deciphering the letter and misses his chance to represent earth at the wedding of a Martian princess. As they say, it's the thought that counts.

[The work and theories of Schiaparelli and Lowell; the science fiction pulp yarns from *Locked City* to *Strictly Formal*; the uncertain prose and format of *Across the Zodiac* to the high adventure of *John Carter of Mars*; even a brass-filled musical toccata entitled *Mars: the Bringer of War*, written in 1914–1917 by Englishman Gustav Holst; this was Mars of the early twentieth century. A world of countless unknowns on which to let the imagination run free. Today, science and literature are considerably different. They have been enlightened by a decade of concentrated, very dramatic discovery, beginning with Mariner IV and holding, at the present time, with plans for the Mars-roving Viking III. We have proven that the ever-present canals of Mars are simply illusions, caused by the way in which our atmosphere melds the Martian craters into thin lines; in brief, we have fostered the sophistication which instinctively rejects fantasy. Since we need the science to better our lot as well as the dreams to keep us one step ahead of the science, we are clearly faced with a desperate paradox! But this is the meat of a later chapter. For in the half-century between Lowell and Mariner, Mars was still a vague world in both feature and composition. We picked up smatterings of knowledge here and there, but nothing so extraordinary as to corrupt our dreams about canals and exotic Martian life forms. Yet, science fiction about Mars did not stand still during this period. Yes, there were the trite John Carter facsimiles, the altruistic aliens, the bug-eyed monsters, and the earth people lost on Mars. But these faded to insignificance beside the literature and motion pictures which *broke from the past, efforts stimulated by events that transpired on earth!* We have already seen how a political situation can affect science fiction in *Visitor to Earth*, which was written during World War II. Now, let's look at how earth of the late forties and the fifties influenced those storytellers who inherited Mars from Wells, Burroughs, et al.

4

GROK

When H. G. Wells wrote *War of the Worlds*, he did more than
look askance at his nation's colonial policies. He created a science fiction parable, and it opened the field to countless such commentaries.
In fact, growing up in a world that became smaller and more vola-
tile by the day, the genre found itself beset by legions of *both* escapist authors and vociferous moral marksmen. To coin a phrase, the
medium had become both a massage and a message! Mars was a
microcosm of this literary blossoming, and authors used the planet
not only to entertain us, but to defrock organized religion, personify
the power of the atom bomb, tell caustic jokes, warn us of a dangerous future, point out the vulgarity of prejudice, and so forth. But while
the novelists—and filmmakers, as well—were getting ready to flex
their imaginations, the scientists, as ever, were busy exercising their
eyes, brains . . . and tongues.
[From the first, there were those who felt that science would be
best served by Percival Lowell if he and his canals were reduced to
but silly footnotes in astronomical history. Others, primarily laymen,
believed that Lowell had done more to fire public interest and advance the exploration of the planets than any man since Galileo.
Perhaps astronomer and Project Viking scientist Carl Sagan (see
chapter six) accurately summed it up in a recent interview: "[Lowell]
triggered the current interest in Mars for, while his flamboyant ideas

69

had turned off the whole contemporary community of astronomers, they turned on all the eight-year-olds who came after him, and who eventually turned into the present generation of astronomers." Why was there such hostility toward Lowell? One would imagine that scientists, so dependent upon grants to pursue their work, would have *welcomed* the publicity he garnered. Unfortunately, Lowell's antagonists, so stoic behind a telescope, were extremely jealous of his celebrity. This aggravated their already strong disapproval of his scientific methods. They felt, and not without justification, that Lowell's alien panorama was strictly a product of the astronomer's fancy and, as such, had no right to be confused with what we really *knew* about Mars. As for the canals themselves, despite Lampland's photographs and Lowell's detailed charts, maps, and globes, there were still only a few astronomers who had actually *seen* these markings. And even to them, in those rare moments when the atmospheres of both Mars and earth were perfectly still, our largest telescopes rendered the planet a hazy sphere the same size as the moon when viewed with the naked eye. Thus, poor visibility compounded the professional envy and an unwillingness by Lowell's peers to embrace a Martian race and its desperation waterways. Of course, through the perspective of hindsight, we find fault on both sides: Lowell had an inquisitive mind, but it was strongly influenced by an eager imagination, while his opponents were pragmatists who were too greatly taken with their own pretention. But Lowell's detractors *did* have tradition on their side: scientists and their theories are invariably guilty until proven innocent. In other words, as far as most astronomers were concerned, there were and would be no canals until somebody up and *drowned* in one! As a result of this friction, the telescopic study of Mars fell to bitter disrepute. Early in the 1920s, astronomers began turning their equipment and astrophysical formulae on the stars, trying to gain a broader picture of our universe and its origin, and leaving the Red Planet largely unattended. What little we *did* learn about Mars was either factual or soundly speculative, but it was all anticlimactic after the wonders that had issued from Flagstaff. A case in point is this very question of life on Mars as it was bandied about between 1910 and 1965. Even while Lowell was devising his grand scheme of alien survival, scientists like the Swedish chemist and Nobel Laureate Svante Arrhenius sought alternate explanations for his observed phenomena. Unfortunately, without the inherent drama of an advanced civilization around which to rally the Martian discoveries, they take on a somewhat clinical air. For example, where Lowell had attributed the once-yearly darkening of the greyish-green areas or *maria* (seas) to receding or flourishing vegetation, Arrhenius decided, in 1912, that the change might simply be a chemical reaction sparked by water from the melting polar ice. This theory retained the vanishing caps while suggesting a credible, nonlife explanation for the consumption of water. Indeed, as recently as 1960, this idea was still acceptable to chemists, with the stipulation that ultra-violet radiation from the sun had some bearing on the change. More recently, M. Audouin Dollfus of the Paris Observatory offered a rationale somewhere between those of Lowell and Arrhenius: "The dark areas," he said, "may consist of tiny particles which undergo changes in shape and size with the seasons." Since minerals don't have this property, the shadowing of the surface was thought to be due to a type of algae or lichen that was polarizing in the presence of the iron ore limonite. Certainly the planet's spectrum supported this theory, ideally complimenting the needs of these simple growths. As for why the change occurs at all, "Their superficial colored pigment," Dollfus concluded, "affords them protection against cold and excessive radiation." Clever—but for all its vaunted sophistication, the theory is *dead wrong*! As we'll see in a moment, the cause of the shading is the movement of Martian dust. However, that doesn't rule out the prospect of finding life on Mars.

[The idea that life could have evolved on another planet was first given route and reason by the Soviet biochemist Alexander Ivanovich Oparin. In 1936, Oparin wrote a book entitled *The Origin of Life on Earth*, in which he hypothesized that organic compounds, such as those we discussed in chapter two, were a product of the sun's rays acting upon various atmospheric gases. Prior to this theory, which gave a chemical basis to the origin of life and dismissed it as a purely terrestrial wonder, scientists had simply assumed that organic matter had always existed. As for there being any of Oparin's building-block substances on Mars, this has always puzzled planetary scientists. W.M. Sinton thought he had found evidence of their existence after examining light reflected by Mars' changing dark areas: the absorption lines (à la Fraunhofer) were similar to those of carbon-hydrogen bonds. Unfortunately, a decade later, *Sinton's Bands* as they were called proved to have been those of hydrogen deuterium oxide in the *earth's* atmosphere. If nothing else, this illustrated the need for out-of-atmosphere devices with which to study the heavens—or a different

71

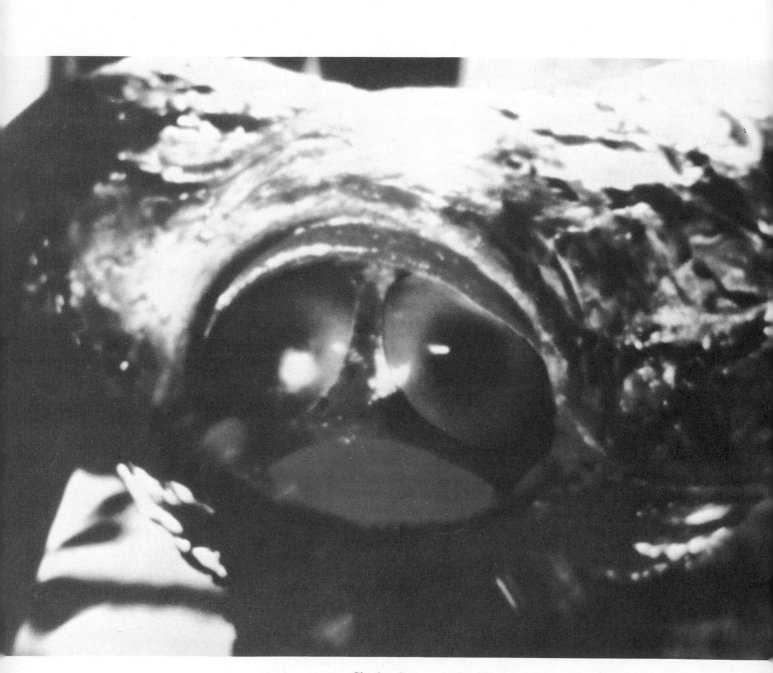

Charles Gemora as the Martian in *War of the Worlds*.

type of earth-based research altogether! And, at the time, only one of these was practical. Thus, in the early 1960s, scientists J.A. Kooistra, R.B. Mitchell, and Hubertus Strughold of the United States Air Force School of Aviation Medicine at Randolph Air Force Base, Texas, devised and performed just such an experiment. Gathering four types of soil from distinctly different geographical locales, they kept the inherent microflora, ground the samples together until they were homogeneous, sterilized and dried a portion of the soil type to serve as a medium, and used the unsterilized portion as an inoculum. The conveyor samples were then mixed with water and the soil inoculum to obtain a desired moisture level of 1 percent. Finally, this mixture was put in a Martian simulator chamber, base-line counts were made, and the sample was exposed to environmental conditions similar to those we might expect to find on Mars. After three months, it was shown that the micropopulation of the soil samples had changed to accommodate those organisms that were capable of surviving and multiplying under the simulated Martian conditions. This hardy population, it was discovered, was composed primarily of facultative anaerobes—versatile bacteria that do not require free oxygen in order to survive. Interesting? Yes. Persuasive? Yes again. Yet, even these findings were inconclusive. They proved that certain microbes can conceivably *survive* on Mars, but did not show that, unprovoked, they would have actually *evolved* on Mars! Lowell, at least, had had the foresight to warn us about forcing terrestrial biology on the search for life on Mars. In fact, as poetic justice would have it, this line of reasoning eventually gave us the key to the changing maria. The Dollfus lichen idea began to buckle in the fifties and sixties when the drab, seasonal plants failed to satisfy the colorful, cyclical changes we saw on Mars. Thus, wisely pursuing an entirely different tack, astronomer and stellar spectroscopist Dean B. McLaughlin advanced, between 1954 and 1956, the volcanic-aeolian theory of the darkening maria. In its original form, it precluded the causation as being necessarily organic, involving other elements of the Martian persona which have long puzzled scientists. McLaughlin speculated that the regular color changes coincided with the simultaneous activity of volcanoes and the blowing of winds caused by the heat of summer.[1] It

1 During their respective summers, the Northern and Southern Martian hemispheres receive 44 percent more radiation from the sun than they do during the winter. Contrarily, the seasonal difference on earth is negligible.

was these winds that carried volcanic ash about the planet, creating shaded patterns on the maria. Or, so as not to exclude the possibility of life, McLaughlin presumed that volcanic ash, with the moisture it brings, might also influence the growth of vegetation that, in turn, results in the darkening. As it developed, Mariner IX proved, in 1971, that while the changes may not be of a volcanic nature—there are a number of active craters on the planet, but none where McLaughlin said they should be; perhaps there are smaller vents of which we are unaware—they *are* the result, not of life, but of wind-shifted dark and light dust.

[It should be obvious by now that this post-Lowell period of Martian observation was constructive and thought-provoking, but utterly lacking in authority. There *was* some hard fact. We studied Mars' surface temperature using thermocouples, and found it to be as high as 80 degrees F. at the equator, and as low as −190 degrees F. at the poles. We analyzed the Martian atmosphere and identified many of its components: in 1947, Gerard P. Kuiper found absorption bands (in the infra-red) that were caused by the presence of carbon dioxide. How *much* carbon dioxide, we didn't know until the 1960s, when it was shown to be the atmosphere's foremost component. We also knew that there was little or no free oxygen on Mars, but were unsure as to the ratio of nitrogen (it comprises four-fifths of the earth's atmosphere and is a major part of the protein compound) with the other elements. But what did these discoveries *mean*? Well, they *did* lead to some interesting notions about the polar caps and the surrounding blue band, for example. The low temperatures in the extreme north and south of the planet, which are below the freezing point of carbon dioxide, inspired the theory that the caps might indeed be frozen continents of the gas. Spectroscopic data argued otherwise. It was then put forth that if nitrogen and oxygen had formed nitrogen dioxide and nitrogen tetroxide, they might also exist together as nitrogen peroxide. Unfortunately, an abundance of this gas would mean a virtually dead world, since it is toxic to all plant life. In any case, since no further reactions are likely between oxygen and nitrogen, the planet's surface was thought to be made up primarily of such decayed matter. Accordingly, it was speculated that the polar caps might be ivory-white deposits of solid nitrogen tetroxide. In warmer weather, these elements were thought to take on a liquid form that was responsible for the blue bands about the caps. And, with the climbing temperature,

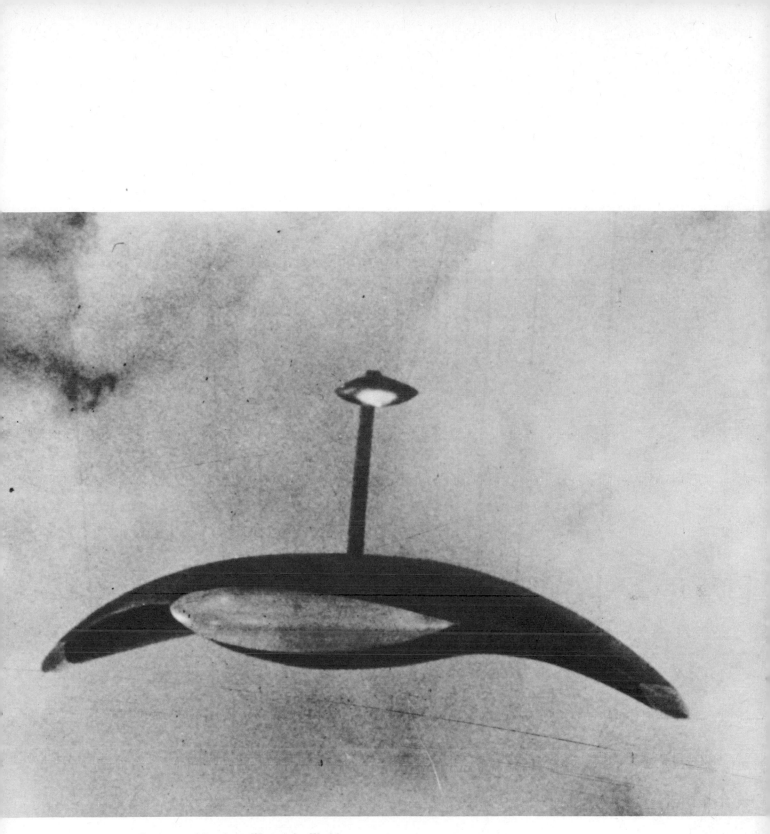

A war machine from *War of the Worlds*.

A war machine levels a gas storage tank in *War of the Worlds*.

the heavy nitrogen tetroxide might be released, moving from the caps to produce that wave of darkening across the maria. This nitrogen-oxygen plan even explained the red color of Mars as being due to the heavy absorption of blue light (thus reflecting red to earth) of nitrogen peroxide. Yet, with all the evidence seeming to support this broad and logical scheme, the presence of these compounds must produce strong absorption bands in the infra-red portion of the spectrum . . . and this is something which Mars does not do. Thus, while scientists had a mass of information at their command, there were few concrete conclusions to be drawn from it. However, this was not a problem for the fantasists, who never disputed Mars data, pro or contra. They simply used it as a springboard for their work, the meat of which was imagination, philosophy, an eye for entertainment . . . and, on one occasion, something more.

[As we have indicated, the bulk of Mars fiction in the mid-twentieth century was either vivid and/or moralistic. However, the crucial question of responsibility was *also* raised with Orson Welles' dangerously slick *War of the Worlds* radiocast. Although it was announced numerous times that the Halloween night, 1938 program was a dramatization, it is clear from the strong documentary air of the show that Welles had had it in mind to alarm at least a *few* members of his audience. An updating of the H.G. Wells tale with a script by Howard Koch, the broadcast balanced the innocuous with the terrifying, cutting between the performance of a typically dreary dance band to the landing of strange cylinders in a field near Princeton, New Jersey. At first, the mood of those listening to the CBS Mercury Theatre presentation was subdued. People were either entertained by the show or, not knowing it was a play, stayed tuned for the latest information. But by the time the aliens loosed their heat rays on people surrounding the cylinders, latecomers had either armed themselves and barricaded their homes, or were fleeing the populated city and town areas entirely! Given the distance of 40 years, this widespread hysteria is far more interesting than the melodrama itself. As a work of emotion and art, the novel is much superior to the radiocast. But a social historian, well versed in both psychology and our space program, would do well to use the invasion as a study resource on how humankind must prepare for eventual contact with extraterrestrials. It is an important inevitability, and one we will examine more closely in chapter seven.

[If we exclude the Welles effort, Mars has traditionally enjoyed its greatest impact in the print and movie media. Yet, while the same creative drafts sway both novelists and filmmakers, the two art forms are greatly dissimilar. For instance, there's a gross disparity in economics. The audience which would make a science fiction novel successful, sales in the tens of thousands of copies, would consign a movie to oblivion. Because of their enormous production expenses, movies must be seen by millions of people. Thus, producers instinctively seek a common entertainment denominator, generally gravitating toward the simple-minded rather than the multileveled. Science fiction films also stumble over their gaudy special effects and brutish monsters, elements which are usually included at the expense of plot and characterization. Therefore, what we will find as we turn first to the over 60 films that have featured Mars and Martians, is colorful pap, isolated works of some merit, or lessons in Lowellian astronomy. After all, what's more interesting to the general public: canals and highly evolved (and malevolent) batrachians, or rolling fields of greyish lichen? In fact, the dust had hardly settled on Lowell's theory of maria vegetation when the 14-year-old moving-picture medium gave us Thomas A. Edison's three-minute-long *A Trip to Mars* (1910), in which a scientist drifts to Mars on wings of anti-gravity dust, meets humanoid Martians, and battles a forest of tree monsters. Three years later came *A Message from Mars*, an hour-long adaptation of an 1899 play by Richard Ganthony, the story of how the God of Mars (not the Olympian War God) comes to a self-centered man in a dream, shows him the pain of his fellow beings, and shames the mortal into doing his share to help the needy. The British film was remade in Hollywood eight years later, but was second-rate Dickens in both versions. *Pawns of Mars* (1915) was a more gadget-laden tale, as earth was given a peek at the advanced science of the Red Planet, while *The Sky Ship* (1917), a film remarkably like Cromie's *A Plunge Into Space*, transports Danish astronauts to Mars where they find an idyllic society and bring a Martian female back to earth. The feature-length *A Trip to Mars* (1920), an Italian effort unrelated to the Edison picture, was considerably less realistic, taking its crew to Mars via airplane; *Radio Mania* (1922)—also known as *Mars Calling*—was the next Martian film, in which a man dreams that citizens of Mars show him how to transmute the elements. However, it wasn't until 1924 that the world was given its first Martian classic, the Russian-made *Aelita*. Although the film is a clear slap at the late and unlamented czarist regime,

Aelita is remembered *not* as the story of how the Empress of Mars (Yulia Solntseva) falls in love with the cosmonaut Los (Nikolai Tseretelli), while his copilot Gusev (Igor Ilinski) incites the Martians to rebel against their autocratic ruler. The silent picture, based on a play by Alexi Tolstoy, is important as the first moving picture to suggest an alien world using design and perspective that was more expressionistic than functional. The Martian workers are represented as slaves by the heavy restrictiveness of their uniforms, while the sets are similarly regimented in hard, geometric patterns. In all, the picture is a fine, stylized evocation of a world different from our own—physically, if not politically. It is also an interesting socio-historical record of the Soviet mind and heart in these post-Revolution years.

[Following the Russians into space—as we would do until the late sixties—the United States came up with the handsomely mounted but—less subjective *Just Imagine* (1930), in which star El Brendel falls asleep for a half-century and wakes up in the year 1980. After admiring the awesome sights of a future-time metropolis, El becomes involved with a flight to Mars from which we learn that Martians come in duplicate, one set being good and the other evil. Perhaps *here* was the explanation for Lowell's double canals! However, *Just Imagine*, like most of the earlier efforts, was a gimmick-filled novelty only modestly received by the public. In fact, Mars didn't make the cinematic big time until it was visited by the supreme hero of space fiction in the commercially successful but artistically destitute *Flash Gordon's Trip to Mars* (1938). The three Flash Gordon serials[2], all of which starred the jut-jawed Buster Crabbe as Flash and Charles Middleton as his vile nemesis Ming the Merciless, were every bit as melodramatic as the pulp science fiction stories, but featured technology of the sort to make Hugo Gernsback's machine oil clot! Flash and Dale Arden (Jean Rogers) had met the brilliant Dr. Zarkov (Frank Shannon) in the first Flash Gordon adventure (1936), when the trio flew spaceward in Shannon's homemade rocket to stop the runaway world of Mongo from colliding with the earth. Instead, they collide with Ming, dictator of the stray planet, who is destroyed after 13 chapters, and our world saved. However, Ming returned in the 15-chapter *Flash Gordon's Trip to Mars*, using the Red Planet as a base from which to draw away the earth's nitrogen. Our hero and his friends hasten to the

2 The serials were 12-15-chapter adventures, with one fifteen- to thirty-minute episode shown every week at theatres across the country. These chapter-plays would either run on a bill with other serials, or accompany the showing of a feature film.

fourth planet, where they battle zombies, clay people, a sorceress, tree people (the children of the arboreal creatures from the Edison film, no doubt), fight their way through the Valley of Desolation, convince the humanoid Martians not to ally themselves with Ming, and eventually force the madman into a disintegration chamber. The chapter titles pretty well sum up the action: *New Worlds to Conquer, The Living Dead, Queen of Magic, Ancient Enemies, The Boomerang, Treemen of Mars, Prisoner of Mongo, The Black Sapphire of Kalu, Symbol of Death, Incense of Forgetfulness, Human Bait, Ming the Merciless, The Miracle of Magic, A Beast at Bay,* and *An Eye for an Eye.* Ming's last act of hostility came in *Flash Gordon Conquers the Universe* (1940), the 12-part story of how Ming comes to earth, contaminates our atmosphere with particles of the Purple Death, and is once more foiled by Flash (he's run over by a spaceship). Fireworks, frenzy, and villainy are the cornerstone of these exciting adventures. However, like Arnold's novel *Lt. Gulliver Jones,* the goings-on are all rather witless. Conversely, like John Carter of Mars, Flash Gordon had an infectious hubris—or *chutzpah*, as they say on Mars—that kept audiences returning week after week to join the space ranger in his dangerous escapades.[3]

[Ming wasn't the only one with a vested interest in Mars and the earth, as we learned in the 15-episode thriller *The Purple Monster Strikes* (1945) and its 12-part sequel *Flying Disc Man from Mars* (1950). Intending to conquer the earth, the Emperor of Mars (John Davidson) dispatches an agent known as the Purple Monster (Roy Barcroft) to steal plans for a spacecraft capable of traveling round-trip between worlds. Arriving on our planet, the scout kills the rocketship's designer, Dr. Cyrus Layton (James Craven) and, entering his body, assumes the scientist's identity. Then, with the help of a criminal named Garrett (Bud Geary), the alien begins construction of the interplanetary vehicle. After several tense weeks in which lawyer Craig Foster (Dennis Moore) and Dr. Layton's niece Sheila (Linda Stirling) try and stop the Martian, they turn an electro-annhilator on the fiend and his completed transport, destroying them both. Five years later, however, the Martians try again as Mota (Gregory Gay) offers aircraft expert Dr. Bryant (James Craven) advanced Martian

3 The three Flash Gordon serials were later edited into featurelength films. *Flash Gordon* became *Rocket Ship, Flash Gordon's Trip to Mars* was renamed *Mars Attacks the World,* and *Flash Gordon Conquers the Universe* was called *Peril from the Planet Mongo.*

knowledge if Bryant, in turn, will help build a fleet of atom-powered planes with which to subjugate the earth. Bryant agrees, and work gets underway. Although Ken Fowler (Walter Reed) and his attractive secretary Helen (Lois Collier) pick up Mota's trail, the alien thwarts them with his thermal disintegrator and a nuclear spaceship—it climbs vertically, hovers, and can fly in any direction and at any angle without turning around. Finally, however, Kent and Helen track Mota to his hideout in a volcanic crater. There are fisticuffs and, during the scuffle, an atom bomb slips into the pit and explodes. While our hero and his secretary are able to escape in one of Bryant's compact, swept-wing jets, Mota and his cronies are killed. As with the Flash Gordon films, we are once again confronted with untaxing entertainment which, if nothing else, underlines the image that both mythology and Lowell have given to the planet Mars.

[Inspired by the newborn atomic age, and given even greater impetus by the study of the awesome German V-2 rockets captured after the war, a new kind of science fiction film made its way to the screen in 1950. The picture was *Rocketship XM*, and it carried Lloyd Bridges and his three-man, one-woman crew to a strangely prophetic Mars of jagged rocks and bleak deserts. There, the explorers find a Martian civilization reduced to savagery in the aftermath of nuclear war. The astronauts are captured by the radiation-scarred barbarians, Bridges and the girl escape, and they blast off for home. Unfortunately, there isn't enough fuel for the return trip, and the fliers die in a crash landing on earth.[4] Plausible situations, supported by rational scientific hardware and solid, low-keyed performances make *Rocketship XM* a valuable record of what both technology and drama agreed might one day be the scenario of a trip to Mars. The picture was a tremendous box-office success and, although marred by a hurried production schedule and low budget, it proved that space flight could be the subject of a thoughtful and adult drama.[5] However, these aesthetic realities were not foremost in the minds of the Hollywood bankers. *Rocketship XM* made money, so the rule of thumb was "let's

4 Although *Rocketship XM* was a black-and-white film, scenes on the surface of Mars were tinted red violet for effect.
5 Actually, *Rocketship XM* was not the first adult space drama to go before the cameras; it was simply the first to be released. When Hollywood saw the publicity George Pal's revolutionary film *Destination Moon* was receiving from the press during its lengthy production, *Rocketship XM* was conceived and filmed in under three weeks, and rushed to theatres before Pal's lavish space epic. As a result, they are almost identical in theme and approach.

do it again." *Flight to Mars* (1951) was the result, with Cameron Mitchell leading an expedition spaceward. The twist here is that the Martians, a sophisticated race living underground and ruled—we kid you not!—by Queen *Alita* (Marguerite Chapman), have made plans to invade the earth. Naturally, the astronauts sabotage Alita's strike force and save our world. But there would be other assaults, many of which came woefully close to succeeding. Not that all the screen's Martians were antagonistic: in 1952, for example, Peter Graves established radio communication with the *Red Planet Mars*, learning that the world is inhabited by God and an advanced civilization of Christians. However, since pacifist aliens are not good box office, the traditional warrior-Martians were brought back a year later—and in a big way.

[*War of the Worlds* became a motion picture in 1953, helmed by George Pal, the dean of science fiction films and producer of such classics as *When Worlds Collide* (1951), *The Time Machine* (1960), and, of course, *Destination Moon*. Like Orson Welles, the filmmaker took several liberties with the story. The pit and the general plot were retained, but Pal changed the locale from nineteenth-century England to contemporary Los Angeles. He felt that the young audience for which his picture was tailored would find it easier to identify with a modern setting than with the world of Wells. The adjustment also allowed Pal to pit twentieth-century technology against the aliens. It's one thing to read about the crushing of a defenseless Great Britain; it's something else entirely to see planes, tanks, and artillery (which were used in the novel) being atomized by the monsters' screeching heat and disintegration rays. Pal also redesigned both the aliens and their fighting machines. The Martians no longer resembled octopuses, but were more like ambulatory slugs; their war vessels were pictured as flying saucers rather than as tripodic titans.[6] Thus, the screen effort which emerged from *War of the Worlds* is a $2-million special-effects Oscar winner, an eyeful of crumbling city blocks and vaporized flesh, and an earful of human misery accompanied by the strident screams of weaponry. The subtler Wellsian allegories are gone; the anach-

6 Pal had originally wanted tripod war machines, and had his technicians wrestle with the problem for several months. Test scenes were shot using domes riding atop columns of electricity. The effect was magnificent. Unfortunately, they were pumping over one million volts of electricity through the models and down along thin wires to the floor. The trick was deemed too dangerous and was abandoned.

A pair of aliens from *Invasion of the Saucer Men.*

Ray Corrigan as the stowaway monster in *It! The Terror from Beyond Space*.

ronism of the metallic giants running amok in a nonmechanized England is no more; but the spectacle and the all-important humbling of our race are intact. One cannot help but feel that if Wells had lived to see the film, he would have graciously approved. Oddly enough, the extraterrestrial stars of *Invaders from Mars* (1953) are more in keeping with Wells' physical descriptions and caste system. Commanded by a small, multitentacled blob, a Martian saucer worked by pop-eyed humanoids buries itself in the sands of a desolate North American field. A young boy (Jimmy Hunt) sees the spacecraft land, but has a difficult time proving it's really there: those who go to investigate are captured by the aliens and have surrogate thought centers imbedded in the base of their brains. These electronic devices allow the Martians to control the earth peoples' every move. In the end, soldiers storm the ship, blow it to pieces, and the menace is ended. At least, for *us*, that is. The Martians, however, might have had their tentacles full if Bud Abbott and Lou Costello had reached their goal in *Abbott and Costello Go to Mars* (1953). Instead, the comedy team's ship speeds off course and lands on Venus. And what does the duo *find* on the Planet of Lovers? An all-female population, of course, played by contestants in the Miss Universe competition! Yet, Mars was not without women: in 1954, earth met Nayah (Patricia Laffen), the *Devil Girl from Mars*. Nayah, with the help of her huge robot Chani, was planet-hopping in search of men to abduct for breeding purposes. Interestingly enough, whereas the Venusian ladies were all scantily clad and of a soft demeanor, Nayah was garbed in black and had a mean, "God of War" disposition. Coincidence, or our subconscious mythological programing talking?

[Having reaped tremendous critical and financial rewards from *War of the Worlds*, George Pal returned to Mars in 1955 for a dramatization of Werner von Braun's original and visionary monograph, *The Mars Project* (see chapter six). This time, however, Pal was several million miles off target. Launched from a space station orbiting one thousand miles above the earth, a five-man rocket is stranded on Mars, and it looks as if everyone is doomed. The planet is an arid, brick-red world where nothing grows—until a seed nurtured by Christmas Day snows breaks the Martian soil. Sustenance follows hard upon, and the pioneers survive until it is feasible to return to earth. They have proven that human beings *can* colonize Mars. Although there are some interesting vistas of the barren world, and a hint of the difficulties in surviving on an hostile planet, *Conquest of Space* is, overall, a dull motion picture. This is due to the fact that until the Martian climax, the film is simply a documentary of life on a space station. Uncharacteristically shoddy technical effects further mar the effort, a star which shines none too brightly in the distinguished Pal repertoire. More entertaining is *World Without End* (1955), in which the Red planet figures only tangentially. Breaking from Mars orbit, a rocket crashes through the time barrier and lands on earth of A.D. 2508. Perhaps this is the resolution to Pal's episodic space saga . . . ?

[The catch phrase "little green men" has always been a popular collective when referring to Martians. Yet, the generic nasties themselves did not reach the screen until 1957, when yard-tall, bug-eyed, bulb-headed aliens landed in the small town of Hicksville in *Invasion of the Saucer Men*. Teenagers Johnny Carter (Steve Terrell) and Jean Hayden (Gloria Costillo) know that the Martians have arrived, but while they try to convince townspeople of the danger, the invaders start killing off the nonbelievers, using their long, hypodermiclike fingers to inject them with the alien blood counterpart, *wood alcohol!* Fortunately, the kids are able to recruit a teenage attack force. They learn that bright light is fatal to the Martians, surround them with cars, turn on the headlamps, and reduce the beings to ash. It's too bad, however, that once again we see the story from the terrestrial point of view. Exploring the Martians' personality, as well as their needs and habits, would have spawned some interest and drama in even a Grade B quickie like this. *It! The Terror from Beyond Space* (1958) came closer to achieving this goal, as a Martian stowaway on the first earthship to Mars makes a shambles of the return voyage. Like Wells' aliens, the tendrilled encroacher (played by Western film great, the late Ray "Crash" Corrigan) subsists on human blood. After a spate of murders, followed by tense, hide-and-seek chases throughout the spacecraft, the crustaceous alien batters its way through walls and metal doors to corner the surviving astronauts in the control room. In a final, desperate maneuver, the spacemen don their pressure suits, open a port, and watch as the air pressure plummets and the creature is sucked into space. Objectively, of course, we feel some measure of sympathy for the monster. Its death is a horrid one, and, after all, the maneater was only trying to get a good meal. As H. G. Wells pointed out, are we any less bloodthirsty in our own butchery of cattle, fish, and fowl for food? Alas, such higher considerations were displaced by a lean-witted Flash Gordon mentality in the subsequent Mars films *Angry Red Planet* (1959), which welcomes a complement of astronauts—three male and one female—with an assortment of silly, wire-operated monsters; *Nebo Zowet (The Stars Call)* (1959), a Russian-made epic in which Soviet cosmonauts rescue an American expedition stranded on Mars; *Space Station K-9* (1961), another effort from the U.S.S.R., this time with the two rival nations joining forces to conquer the Red Planet; and *A Martian in Paris* (1961), who has come to earth to gain a better understanding of the uniquely terrestrial emotion called love.

[A mixed bag of Mars movies were made during the early sixties, harking back to mythology in *Mars: God of War* (1962), as the sons of the demigod Hercules battle Olympian terrors; bringing the Martians Ogg and Zogg to earth to steal the plans for a scientist's miraculous flying submarine-tank, only to be bilked and baffled by *The Three Stooges in Orbit* (1962); having the Martians use radio waves transmitted by a Viking-like craft which lands on their world to ride to earth, duplicate people they meet, and successfully take over the world in *The Day Mars Invaded Earth* (1963); and sending a two-man craft to Mars in *Robinson Crusoe on Mars* (1964). Indeed, this last motion picture is the film that *Conquest of Space* only *should* have been. When an asteroid damages the orbiting space vehicle, it crashes on Mars killing one of the occupants (Adam West). The second astronaut (Paul Mantee) is able to struggle from the wreck of the ship and finds himself on a crusty, mountainous world with an almost intolerably thin atmosphere and littered with small craters which shoot fireballs into the green Martian sky. The survivor is frightened and dangerously low on natural resources, yet he does not buckle beneath his lot. By burning porous stones, he is able to fill his air tanks with oxygen released by the combustion; he finds small rivers and edible plants in catacombs beneath the planet's surface. But—after such a fine start—by trying to shadow the plot of the Daniel Defoe classic on which it is based, the picture is almost its own undoing. The introduction of Friday (Vic Lundin), a slave escaped from the flying saucers of his cruel overlords—who are never shown on screen—threatens to reduce the story of courage, loneliness, and discovery to a trite tale of flight and derring-do in the Flash Gordon tradition. However, the viewer is forced to forgive the senseless detour by the heroes' climac-

87

tic trek to the awesome polar caps. A visualization of the volcanic-aeolian hypothesis, as well as a panorama of glacial ice, are well worth the price of admission!

[Filmed on the rugged slopes of Death Valley, *Robinson Crusoe on Mars* tells us as much about the human animal as it does about its stated subject. And that's what excellent motion picture making is all about. Unfortunately, as was the case with this film, product integrity doesn't always pay off at the box office. By cutting a monster from the story, and concentrating on Mantee's problems and appendant mental decay, the picture proved too sophisticated for the average moviegoer, and did not show a profit. The result? It was followed by a baker's dozen of featherweight Mars movies, none of which made an appreciable contribution to the genre. *Martians Have Twelve Hands* (1964) was a Spanish/Italian film in which aliens who come to study the earth find it so appealing that they decide to stay; less amicable were the inhabitants of *Il Disco Volante (The Flying Saucer*, also known as *The Martians*, 1964), a Dino de Laurentiis presentation about Martians who borrow a few earth people for a trip to Mars. Then there were four well-beloved fantasy staples given a run through the space mill: *Wizard of Mars* (1964), a retelling of *Wizard of Oz*, with an ancient Martian race doubling for the citizens of L. Frank Baum's Emerald City; *Santa Claus Conquers the Martians* (1964), a deed he works after being kidnapped to service Martian youths; *Frankenstein Meets the Space Monster* (1965), the story of robot astronaut Frank Reynolds (Robert Reilly) who, after being badly disfigured in a faulty reentry, combats a Martian Queen, her bald-headed, pointy-eared subjects, and her pet monster Mull, all of whom are spiriting away earth women to bear children for the dwindling Red Planet race; and finally *Pinocchio in Outer Space* (1965), in which the boy is turned back into a wooden puppet for misbehaving, visits Mars, and saves the earth from a giant space whale. There were also a number of nonadapted and unastounding Red Planet films: the twenty-first-century *War of the Planets* (1965), sparked by energy beings from Mars; *Mars Needs Women* (1966), featuring former Disney child actor Tommy Kirk as the leader of yet another "snatch-up-the-ladies" expedition; *Thunderbirds are Go!* (1966), starring marionettes as the first humans on Mars, where they discover such life forms as fire-breathing rocks; *Santo vs. the Martian Invasion* (1966), pitting the masked Mexican super-hero against an alien army; *Queen of Blood* (1966) with Florence Marly as

The surface of Mars as seen in *Angry Red Planet*. It is interesting that while the men must wear protective suits, the woman is able to weather the Martian landscape unprotected!

the green, egg-laying, vampiric alien Velona, who is brought to earth when astronauts discover her disabled spaceship on Mars; *Ne Jouez Pas Avec Les Martiens (Don't Play with the Martians)* (1967), the warning to a reporter who fabricates a story about Martians arriving on earth, only to learn that aliens *really are here*; and *We Are Martians* (1967), a Soviet production in which Mars tries to make contact with the earth. While a few of these films are not without some merit, most of them are generally low-budget features with anthropoid Martians and little respect for what science knew about Mars.[7] For this reason, *Five Million Years to Earth* (also known as *Quatermass and the Pit*) (1967) is a welcome change of pace. Digging a new subway system in London, workmen uncover a long-buried spaceship, which they bring to the attention of the noted Professor Quatermass (Andrew Keir). Quatermass, as the viewer who's done his homework is aware, has had previous experience with such oddities in *The Creeping Unknown* (1956) and *Enemy from Space* (1957). Venturing inside, the scientist finds the remains of child-sized, locustlike creatures that have apparently been inside this craft for eons. Eventually, Quatermass learns that the ship is cellularly alive, an interplanetary transport that was literally powered by the *thoughts* of these beings. Deciding to tap the craft's memory, the professor is able to see the ancestors of his Acrididaeic *find as they were when alive and living on Mars*. The Red Planet, it seems, was once more a dying world, dry and overcrowded. However, the aliens turned to earth for *more* than just fresh resources: the third planet offered a natural and comparatively dull-witted population. If any of these creatures could be made to take commands, it would provide the Martians with a ready source of slave labor. So they came to earth, tampered with the brains of our primate ancestors, and in so doing created man. And, while this action came too late to save the Martians, it did leave a semienlightened species in charge of the planet. But there is *more* to this incredible history, as Quatermass discovers that in silhouette, the Martians resemble the horned satan of legend. The scientist reasons that stray thought waves from the ship must therefore have been

---

7 *Thunderbirds Are Go!* had to be accurate since kids, more so than adults, are quick to spot flaws in the conception or execution of science fiction films! How many times have we heard youngsters in a theatre audience cry out, "Aw, that's fake!" or "That's not the way it really is on Mars!"

responsible for projecting alien features about our planet, thus giving birth to the devilish icon. Unfortunately, draining the ship causes this stored power to flood forth, destroying everything in its path. Only after one of Quatermass' associates grounds the discharge with a crane are the Martians laid to rest once and for all.

[*Five Million Years to Earth* is an entertaining mindful, and it presented the three most recent Mars films with a difficult act to follow. *Mission Mars* (1968) makes a decent accounting of itself, the story of living balls of light encountered by the first expedition to Mars; *The Astronaut* (1971) is a different sort of story than any we've come across, a made-for-television movie about the space administration's cover-up of the mysterious death of the first man to walk on Mars. And finally, there is a sensitive Canadian effort entitled *The Christmas Martian* (1971), about a group of children who find a Martian hiding in a Canadian wood just before Christmas. In addition to these feature-length films, there have also been a number of short Martian subjects through the years. Among the live-action works: *From Mars to Munich* (1925), in which an invisible Martian takes a liking to beer and secrets himself in a brewery; an episode of the TV series *Outer Limits* (1962) wherein Martian psychologists Phobos and Deimos come to earth and study a human murder; and, of course, the popular television comedy *My Favorite Martian*, starring Ray Walston as an antennaed, superpowered Martian who comes to live with earthling Bill Bixby. On the cartoon front, Oswald Rabbit battled Martians in the animated 1931 adventure *Mars*; the little-known character Scrappy took a *Trip to Mars* in 1938; and Popeye the Sailor rode a *Rocket to Mars* in 1946, ate his spinach, and pounded the metal of an alien invasion fleet into a Martian amusement park. There was also *Martians Come Back* in 1956; a scientist fighting the God Mars for the secret of nuclear energy in the Hungarian-made *Duel* (1959); and a visit by Martians in *What on Earth!* (1966), a ten-minute-long cartoon produced by the National Film Board of Canada in which aliens mistake cars for the earth's indigenous life form, and human beings as parasites which inhabit them. This picture is particularly interesting in light of the recent puzzlement over the interpretation of Viking discoveries (see chapter six). And, as for multiple confrontations with the fourth planet, Woody Woodpecker met the *Woodpecker from Mars* in 1956, repaying the visit a year later in *Round Trip to Mars*, while the inimitable Bugs Bunny proved to be too much for a Martian gladiator (who wore sneakers in lieu of sandals) in *Hare Way to the Stars* (1958), thereafter becoming *Mad as a Mars Hare* (1963) after stowing away on a Mars shot and being transformed into a Neanderthal rabbit.

[From Thomas Edison to Bugs Bunny, Mars had had, as promised, a spotty history in film. The peaks have been few, the valleys many, and the sum total of the experience diverting but generally empty. With the questions and wonder posed by the planet, its moons, and its disputed canals, as well as the versatility of the film medium, one would have thought that the Martian movies could have mustered up greater substance! Unfortunately, the theory of that "common entertainment denominator" was *not* to serve a full-course movie meal, with something for every taste, but to feed the public bread and water. It is universally digestible, and the hungry are not inclined to be particular. However, Mars did receive a fairer shake in print, since those thousands of book-buying science fiction aficionados *do not* tolerate anything less than a mental and emotional banquet!

[By the fifties, the science fiction books and magazines had become virtual partners in storytelling. Upcoming novels were serialized the pulps; short pulp stories were expanded into novels. The lush writing style of the late 1800s was slowly replaced in the 1920s, thirties, and forties with a less florid, more technical form, while plots were created to emphasize people as well as futuristic hardware. This period also bred the first of the contemporary "name" science fiction scribes, such as Arthur C. Clarke, Robert Heinlein, Issac Asimov, Ben Bova, and others. And sooner or later, each of them turned to Mars. In terms of the postwar, pre-Mariner Mars, authors were still free to create any kind of planet they wished, with incredible civilizations and canals, lifeless plains, or settings that were entirely original. Mars could be approached as a vehicle for satire, political expression, moral observation, cycnicism, or unobtrusive entertainment—just so long as a writer did not insult genre buffs by mocking the field or any of its many parts.[8] Let's sample this variety.

[One of the works which served as a transition between the Gernsback era and the atom age in Mars fiction was John Wyndham's *Planet Plane*, currently known by its reprint title of *Stowaway to Mars*.

8 Because fans of science fiction novels, gothics, westerns, mysteries, etc., are so utterly dedicated to these fields, the bulk of this material is written by a relatively small pool of writers skilled in the fields.

Mars as seen from Phobos in a sequence from *Nebo Zowet*.
This film was seen in America as *Battle Beyond the Sun*.

The Three Stooges try to prevent the Martians from turning a stolen airship's powerful disintegration ray on Disneyland. From *The Three Stooges in Orbit.*

First published in 1935, the novel offers two exciting propositions about Mars: an international contest to send the first manned rockets to the planet, and the quest of a young woman to prove that her late scientist father had, in fact, discovered a robot explorer from Mars roaming about the British countryside. Thus, Joan Shirning sneaks onboard Englishman Dale Curtance's ship the *Gloria Mundi* and joins its five-man crew on their 74-day flight to Mars. Wyndham uses the journey to sketch his characters, after which he sets them down on a world of reddish sand, water-filled canals with low-lying bushes "athwart" them ("bulbous, olive-brown plants not unlike spineless cacti"), light winds, low gravity . . . and armies of machines like the one Professor Shirning claimed to have found years before. However, there is no time for Joan to gloat. With six legs, four tentacles, and bodies of different geometric shapes, the robots are able to capture the girl and bring her to the ancient city of Hanno, where she is turned over for study to a young human male named Vaygan. He is the only native allowed to see her, lest the girl harbor some rare germ from earth. Vaygan, with reddish skin and dressed "only in a pair of kilted shorts made from some gleaming material," is the perfect host. He teaches her the language and tells her the history of his world, describing a time when Mars was not a barren planet. Once there were animals, bright flowers, trees, and seasons of snow and falling leaves. There were also wars, petty nations, and all the foibles of contemporary earth people. Then, as the planet grew older, the water began to vanish. The countries banded together to build the canals, but it has only postponed the inevitable. Thus, the Martians-designed robots that can reason, act, repair themselves, and even reproduce; they were built to inherit the planet, a transfer which was even now in progress. Meanwhile, a second earthship sets down near the *Gloria Mundi*, the Soviet Union's *Tovaritch*. A scouting party visits the English craft and its leader, Commissar Karaminoff, informs Curtance that they intend to make Mars a satellite of Moscow. Naturally, the British are opposed to this, having intended that the Red Planet be an independent part of their own empire.[9] The Russians and the British exchange nationalistic insults until the robot Martians render the

9 There is no reference, in any of H.G. Wells' public papers, to what he may have thought about Wyndham's nationalism in *Planet Plane*. If Wells *did* read it, we can only imagine what his reaction might have been!

dispute academic by tearing the wicked Soviet ship assunder. Shortly thereafter, the robots decide that Joan's free spirit and growing love for Vaygan are a threat to their coming rule. She is returned to the *Gloria Mundi* and the British adventurers blast off for earth. Upon their return, the men are feted, but Joan's fate is rather more tragic: no one believes her story of a human civilization on Mars. Like her father, she is scorned, forced to go into seclusion, and dies six months later—ironically, after giving birth to Vaygan's child.[10]

[*Planet Plane* is primarily a soapbox for Wyndham's commentaries on British and Soviet socio-political views, humankind's relationship with both men and machines, and the need for people to seek challenge and an active role in the destiny of society. This philosophy of committment is especially well delineated in Vaygan's discourse on how Martian lethargy is as much to blame for their fate as are environmental circumstances. Wyndham also takes care to exploit the Mars of science and literature, even having his characters discuss the possibility of finding the worlds described by Burroughs or Wells. Thus, while the novel is strongly opinionated, it is also a textbook use of Mars for the expression of social, popular, and personal convictions.

[Skipping some years along in the twentieth century to 1950, we come to one of the most popular Martian books ever written, Ray Bradbury's *The Martian Chronicles*. A string of 26 stories, the work tells of humankind's association with Mars from the years 1999–2026, covering the conquest of Mars and the disappearance of its natives, the planet's colonization and abandonment, the fall of the earth, and a final return to Mars by two families fleeing their war-torn home world. The vignettes vary in timbre from the humorous *The Earth Men*, as the first astronauts to successfully land on the Red Planet are locked in an insane asylum by the slightly nutty Martian, Mr. XXX; to *Way in the Middle of the Air*, which speaks bluntly and commendably against prejudice. Not that *The Martian Chronicles* is simply one idealistic lecture after another. Whatever the excerpt, Bradbury has written it with poetry, a fine eye for justice, and the guiding maxim that human dignity must always triumph over anything that is crude and mindless. Bradbury has also lavished considerable attention on

10 Wyndham had intended to write the adventures of Vaygan and Joan's child on earth, but, if he did, the novel has never appeared in this country.

Paul Mantee as *Robinson Crusoe on Mars*.

Santa Claus and a pair of earth children greet the Martians inside
their interplanetary vessel. From *Santa Claus Conquers the Martians*.

his locales, rendering pre-earth-colonization Mars and its people with an airy beauty: the natives of the planet, which they call Tyrr, are short, tan-skinned, telepathic, and yellow-eyed, speak in "soft musical voices," sleep on mist, fly flame birds for transportation, keep carnivorous plants in cages, eat crystal buns, cook meat in bubbling pools of silver lava, and, of course, live on the shores of great canals. And, while this vision is more an exercise in the romantic than a serious interpretation of scientific data about Mars—of which, incidentally, Bradbury seems painfully ignorant or disdainful—critical readers will tolerate the author's self-indulgence just to get to the meat of his thoughts. Compared, then, to the Wyndham effort, both of which use Mars to explore the human psyche, Bradbury's opus is hopeful rather than nihilistic, and written on the heart rather than carved in technical or ideological steel. Like our old pulp friends *Locked City* and *A Martian Odyssey*, they represent opposite ends of the conceptual spectrum.

[If we speak of Wyndham and Bradbury as being to the right and left of narrative center, then Fredric Brown's *Martians, Go Home* (1955) is off in another dimension altogether. It is, without a doubt, the oddest Mars novel ever written. The year is 1964, and over one billion Martians descend upon our planet via mental teleportation or *kwimming*. An average two and one-half feet tall, the bald olive-green, 12-fingered Martians come for seemingly no other reason than to pester humankind. They are opaque but intangible, insult anyone within earshot, cannot be intimidated by being ignored, refer to all males as "Mack" and all females as "Toots" ("Why bother to learn a new name for every person you speak to?"), turn funerals to farces by correcting "errors" in the eulogies, make public all top-secret government projects and papers, and disrupt everything from radio broadcasts to sexual intercourse—calling to bear their X-ray vision and night vision when couples turn out the lights and hide beneath their covers. Yet, throughout the novel, the Martians never explain why they have come to earth or when they intend to leave. Various explanations are offered by scientists, politicians, philosophers, and such: that they are devils punishing our race; are trying to discourage us from ever visiting Mars, afraid that when we master space flight, we'll ruin their world as we have our own; or are really altruistic beings working to end the Cold War by removing international secrets and giving the world a common nemesis against which to unite. Ultimately, however, Brown tells us that they were accidents, beings made corporeal by the subconscious mind of science fiction writer Luke Devereaux. This realization, followed by Devereaux's disavowal of their existence, ends the Martians' 146-day reign on earth.

[Earlier, we spoke of the Martians-on-Earth theme as being a favorite of science fiction humorists, and in this category, *Martians, Go Home* is one of our finest novels. The aliens grind our science, culture, traditions, biology, and emotions to fine powder, then stir up a wind to humiliate us further, and there is nothing that humankind can do to stop the obnoxious creatures! The result is a tale of beautifully sculpted insanity, and it could only have been built within the genre which *tolerates* little green men from Mars. . . .

[During the 20-year period between *Planet Plane* and *Martians, Go Home*, Mars was featured as an earth colony in Arthur C. Clarke's *Earthlight* (1955), as the setting for antiterrestrial bigotry in Robert Heinlein's short story *Jerry Was a Man* (1947), as the stage for the first disastrous dinner shared by a Karterian (earthman) and the Zumberian (Martian) Mayor Aardvark of Eastern Canalopolis in the anonymous *Behind the Ate Ball* (anthologized as *A Martian Oddity*) (1950), and so forth. Even Fredric Brown returned to the Red Planet in his serious 1957 effort, *Rogue in Space*, as a battleground for Crag, an outerspace smuggler, and the thinking planetoid of the title. However, the next Martian milestone would not be reached until 1961, with the publication of Heinlein's perennial science fiction bestseller *Stranger in a Strange Land*.

[Valentine Michael Smith was found on Mars by the second manned voyage to the planet. Smith's parents had been on the first such expedition, none of whose eight members survived. In the meantime, the Martians raised Smith and, for all but appearances, had made him one of them. This becomes amazingly clear after the young hybrid is brought back to earth. When the government hires an actor to double for Smith, newspaperman Ben Caxton and nurse Gillian Boardman suspect it's to kill Smith and avoid paying royalties due his mother for having discovered a widely used form of space propulsion. Gillian spirits Smith from the hospital where he is being detained and, without realizing it, becomes precious to the pseudo-earthling: she has innocently given him a drink of water, which is a solemn rite on Mars and makes them water brothers. Thus, when the police come to fetch Jill and her charge at Ben's apartment, Smith protects the girl by touching the officers and causing them to vanish. Then, acting on Ben's suggestion, Jill hastens to the Pennsylvania estate of lawyer, doctor, and best-selling author Jubal E. Harshaw, who hides the delinquent pair. Eventually, Jubal is able to use his connections and set things aright, simultaneously making a worldly and wealthy man out of Smith. For his part, the Mars-trained earthman, dissatisfied with the religions of our world, establishes a church where everybody who joins becomes a sharing, loving water brother who is physically, mentally, and emotionally married to every other church member, and where the ultimate goal is to *grok* . . . to understand all in fullness. Doing this, Smith contends, everyone will become a part of everything, and, thus, everyone will be God. Unfortunately, people beyond Smith's circle of brethren do not take to grokking, and brand the Man from Mars a false messiah. In the end, Smith presents himself and his Truth to a crazed mob, who bludgeon and shoot the spiritual leader. Smith willingly dies—or, as they say on Mars, *discorporates*—but it is not the end of the church. His water brothers keep the faith alive while, elevated to the all-grokking status of an Old One (the spirit of the discorporate being), Smith faces the terrestrial challenge from his new position, that of the Archangel Michael.

[Although Heinlein's characters, situations, and dialogue are thoroughly enchanting, the real joy of *Stranger in a Strange Land* is the detail of its Martian culture, and the way in which it is contrasted with that of the earth. To the huge—"[like] ice boats under sail"— three-legged, egg-laying, long-lived Martians, acting "rightly" at a critical cusp is everything. To perform correctly at these junctions— when contemplation must bring forth a right action in order to permit further growth—reflects well on the Martian; it means that the situation has been grokked in its every possible implication. This holds true whether the deed is destroying a planet—the Martians long ago eliminated the "wrong" fifth planet, thus creating the asteroid belt—or answering a question posed by another Martian, something which might require centuries of reflection to do. Eternity, to a Martian, has no meaning. And when something *has* been grokked, it is usually expressed as a work of art. Indeed, so intense is this creative act that, once, a Martian had discorporated while thus expressing himself and *didn't even realize it!* These, then, are the values which Valentine Michael Smith brings to earth, along with the Martian power to freeze time, control his body absolutely, separate spirit from physical form, discorporate people, communicate telepathically, and manipulate objects by telekenesis. About the only attitude which Jubal and his people ask Smith to discard is his cannibalism. On Mars, to have one's body eaten is considered a cherished form of grokking and a great honor.

[The critical difference between Heinlein's earth and Mars is the logical and exhaustive thought patterns of the Martians as opposed to the impulsive actions of humankind. Smith, of course, finds that by joining the two he not only groks like a Martian, but is able to share and grok with others, exploring them through sex and emotion. This is something the Martians are unable to do. Perhaps, in reality, such mutual satisfaction is impossibly idyllic; humankind is too close to the beast in his past to purge life of jealousy, greed, fear, and hatred. However, after observing the peace and fulfillment of Smith's water brothers, as opposed to the pain and stupidity of the two worlds which surround them, one is compelled to acknowledge the sense of Heinlein's vision.[11]

[Two years after the publication of *Stranger in a Strange Land*, Heinlein returned to the Red Planet with *Podkayne of Mars*, the adventures of Marsgirl Podkayne Fries on earth and Venus. Light reading for the teenage market, *Podkayne of Mars* has little to offer the historian of the fictional Mars. On the other hand, British author Michael Moorcock gave us a splendid Burroughs-tradition adventure in the 1965 Michael Kane of Mars trilogy. Sent to the Red Planet of eons past by a malfunctioning matter transmitter, Kane finds that Mars was not always the dead world that it is today. In the first novel, *Warriors of Mars*, Kane is dropped through a purple sky onto a field of red ferns and yellow grasses, where he meets the ravishing Shizala, Queen of the Kanala. She leads Kane to the city of Varnal, along whose graceful white spires creep vines of green mist from a central hot lake. With the aid of a telepathic necklace, the scientist and semiprofessional swordsman learns from his hostess that he is on Mars—Vashu, in the vernacular—and that she is a *Brahdinaka*, or

11 For completists, Heinlein does mention the Martian canals, noting that they were, indeed, real and were bordered by the ruins of ancient Martian cities.

Tommy Kirk points a ray gun at an underling in *Mars Needs Women*.

Princess. Kane explains that he is from earth, although Shizala is skeptical: Negalu, according to reports sent back by *akashasard* (ethercrafts) is a planet of steamy jungles and huge lizards. It is then Kane realizes that the matter transmitter sent his whirling through time as well as space. Later, he would learn that the people of Vashu eventually migrate to Negalu, as the drying of their native world threatens them with extinction. Thus does Moorcock account for the Sumerians! However, a crumbling environment is not the Kanala's most immediate worry. Shortly after Kane's arrival, thousands of *Argzoon*, the Blue Giants of Vashu, come galloping against Varnal astride huge *daharas*—ape-like beasts with four legs and a kangaroo tail. The attack lasts for days, during which time Kane assaults the army of eight-foot-tall barbarians with a flying machine, a remnant of the long-dead and legendary Sheev civilization. Unfortunately, Shizala accompanies him and, when the newcomer lands to do battle, she is taken prisoner. Reinforcements arrive from other nations and the Argzoon are dispersed, leaving Kane and the Brahdinaka's brother Darnad free to find Shizala. After bouts with cutthroats, monsters such as the two-headed, eight-legged, yellow-eyed *heela*, and finally Horguhl, Queen of the Argzoon, the Mistress of Varnal is rescued. However, as she and Kane make plans to wed, the earthman's aides manage to repair the matter transmitter and bring the scientist home. No one believes his story, and on the orders of physicians and psychiatrists, the heartbroken warrior is not permitted near the teleporter. Thus, he builds one on his own and returns to Vashu in *Blades of Mars.*

[Arriving on Mars beneath the dull glow of the two moons Urnoo (Phobos) and Garhoo, Kane is set upon by a *rhadari,* a multicolored cyclopean bear with a huge crest and spine of matted fur. He slays the creature with the help of the kind Blue Giant Hool Haji, heir to the throne of the peaceful Argzoon counterpart the *Mendishari.* Together, they ride to the village of Asde-Trahi, seat of a planned rebellion against the Priosa, a mystical people who have subjugated the Mendishari. However, Asde-Trahi is invaded by Priosae, and after a futile struggle, the two patriots are forced to flee the butchered and flaming city. They cross a great Martian desert, and with materials found in a ruined Sheev laboratory, build a balloon for a dramatic and strategic return to their land. Unfortunately, winds known as the Roaring Death strike suddenly, and bear the heroes to the Western Continent, once the site of an atomic war, and now a wasteland of mutated plants and animals. However, the balloonists must land here to effect minor repairs on their craft. During this stay, Hool Haji leaves his companion to get logs for ballast. When he doesn't return, the earthman traces his trail to the city of the Shasazheen—huge, articulate spiders. Kane rescues the Blue Giant, they fly to the Mendishari capital of Mendishar and, there lead the people against the Priosa. Hool Haji assumes his rightful rule while, flying south to Varnal, Kane must once again fight Horguhl, who has hypnotized Kanala's allies to make war on Shizala's people. And, although Kane stops the witch and marries his Brahdinaka, he is not fated to enjoy peace. In the final book of the saga, *Barbarians of Mars*, the Brahdi's honeymoon is followed by a trip with Hool Haji to the Sheev tower, to procure some of their scientific instruments. En route, they pass through Cend-Amrid, a city stricken with the virulent Green Death, a disease which turns human beings into zombies. To end this plague, caused by the opening of a Sheev cannister containing germs for biological warfare, Kane and Haji fly North to the tower in search of a possible cure. However, upon their arrival, the heroes find that the place is being looted by barbarians from Bagarad, a nation beyond the Western Sea, near the land of the Shasazheen. The hairy devils capture Kane and the Blue Giant, but the party is itself taken prisoner by a contingent of the powerful Dog-men of Hahg, human beings with the heads of bloodhounds. Left to be judged by the harpy-like Jihadoo, masters of the Dog-men, the prisoners are secretly handed arms by an enslaved Cat-girl. After a frantic battle, the Jihadoo are slain. Kane and Haji throw in their lot with the Cat-men, only to find that the Sheev laboratories and libraries have been destroyed by the marauding Dog-men. Kane and Haji return quickly to Varnal to organize an evacuation before the plague arrives. Meanwhile, Kane accidentally uncovers an antitoxin amidst vials taken from the plundered Sheev stores. Normalcy is restored, and, to date, Kane and Shizala have lived happily ever after.

[Although Moorcock's books have obviously borrowed both the heft and the format of the Burroughs novels, they have a sentimentality that is not to be found in the lively, artfully constructed, but emotionally shallow tales of Barsoom. Moorcock even delves into the philosophy of the various Martian races ("fatalistic . . . with a strong moral code that saves them from decadence"), as well as the psyche of its people. Thus, as literature, while each series has the strengths and weaknesses inherent to the adventure genre, Moorcock has built a fictional Mars which holds its own against that of the Burroughs books. It is only because Burroughs more or less fathered the field that Moorcock's work, as well as any similar efforts, have always been prejudged as spiritless imitations.

[Moorcock was one of the last pre-Mariner authors to tackle the Red Planet. Yet, when the photographic space probe revealed Mars to be a harsh, canalless, and boulder-strewn world, the alluring planet did not disappear from our literature. It simply matured. As we've seen, escapist science fiction was less prevalent in the fifties than it had been in the early years of the mode. But the sixties was an especially unsettling era for the United States, with domestic riots, assassinations, a bitter war in Southeast Asia, and growing antagonism toward the hazy bounds of governmental power. As a result, science fiction novelists often used Mars to capsulize the feelings of frustration which were unique to these midcentury years. For example, Norman Spinrad's *Agent of Chaos* (1972) pits a small band of freedom fighters against a cruel dictatorship, as the Democratic League shuttles between earth, Phobos, Deimos, and Mars to break the stranglehold of the Mars-based, system-wide government the Hegemony. In *Farewell, Earth's Bliss* (1966), author D.G. Compton's concern is more social than political, as Mars becomes the dumping ground for earth criminals, a colony in which one's funamental goal is simply to retain his identity and *survive!* Both novels do more than exploit the planet in the name of humanity. They prove that Mars has a future in fiction, and not just a revered past. Science may have deprived us of the quaintly outdated worlds of Burroughs, Wyndham, Moorcock, et al, but the new breed of writers, with new data in hand, is creating a different kind of Mars novel. This, in much the same way that science fiction writers have always used new information to create ever more eloquent plots and characters. True, in future times, there may be no more thoat or dahara riding, and perhaps we will have more worldly oratory and less heart-tugging, wisecracking chatter than before. But with space exploration now a reality, the human imagination, that element which first inspired the Martian scribes, will be driven to greater romantic and speculative heights than ever.

106

5

RIME OF THE MARTIAN MARINER

Although he didn't know it at the time, Isaac Newton had laid down the fundamental principle of outer space exploration when he said, "To every action there is an equal and opposite reaction." In other words, create some form of combustion within an open-ended cone, and that force will raise the container from the ground. The only prerequisite is that the opposite reaction, or *thrust*, have more push than gravity has pull. However, as early rocket men realized, more important than sending a projectile skyward is the fact that the launching need not be an end in itself. Saddle the rocket with a payload and it becomes the key to humankind's survival.

[　　But we get ahead of ourselves.

[　　Rockets have been with us for over seven centuries, the first such missiles being small cylinders filled with gunpowder and used by the Chinese for a variety of purposes. Of course, no one living in the 1300s knew just how these "arrows of flying fire" worked, nor was scientific theory deemed to be of particular importance. The rockets flew, and that was all that mattered. They carried fireworks aloft, were employed as tools of warfare, and even enjoyed brief service as boosters for our first would-be astronaut, the indomitable Chinese nobleman Wan Hoo. Mr. Hoo was convinced that, with the force of 47 rockets at his back, he would be raised into the air where he could remain aloft, suspended from a pair of kites. Forty-seven peasants

stood ready to light the rocket fuses simultaneously; Wan Hoo gripped
a rigging slung between the kites and gave the word for ignition.
Moments later, he and several coolies were blown to pieces. For a
long time thereafter, through several European wars as well as the
War of 1812, rockets remained primarily a tool of the military. Then, in
the early 1880s, a deaf Soviet schoolteacher read a few books, took
drafting tools in hand, and made an art out of rocketry.

[Konstantin Eduardovich Tsiolkovski was raised in the Kaluga
Province and was inspired to explore rocketry by Jules Verne. "He
directed my thoughts along certain channels. Then came a desire and
the work of the mind." The self-taught Tsiolkovski is principally
remembered for having designed multistage rockets or "rocket trains"
as he called them, ships that discarded drained booster engines to
lighten the skyward-bound load, drawing up plans for all manner of
nozzles and combustion cells to make these rockets work. He also
conceived of using liquid fuels rather than powder as a propellant,
since they were subject to greater exhaust speeds and, thus, promised
more substantial thrust. Not surprisingly, no one in a position of
authority in either scientific or political circles took Tsiolkovski's
research seriously. As was later to be the case with Lowell's critics,
innovation—sound or ludicrous—is usually scorned by those who did
not do the innovating. But, up until his death in 1935, Tsiolkovski kept
at his work, publishing numerous scientific papers, writing science
fiction stories, and always maintaining his strong faith in the fact that
rockets would play a vital role in both our future and the future of
technology. His one great regret was never having been able to
actually build and launch such a rocket, which is where Robert
Hutchings Goddard picked up the fire.

[Born in Worcester, Massachusetts in 1882, Goddard, like Tsiolkovski, was weaned on the writings of Verne and Wells, eventually
transferring his interest from science fiction to science, and receiving
a doctorate in physics from Clark University. Independently of Tsiolkovski, Goddard concluded that fuels such as liquid oxygen and
liquid hydrogen would make the most satisfactory rocket propellants,
and elaborated on this belief in a 1919 paper. In this same report, he
also suggested that rockets thus powered could effectively maneuver
in space. Naturally, the experts discarded this revolutionary claim as
frivolous, such distinguished journals as *The New York Times* going
so far as to state that Goddard was out of his mind, that a rocket could
never function in a vacuum. As it turned out, of course, Goddard was

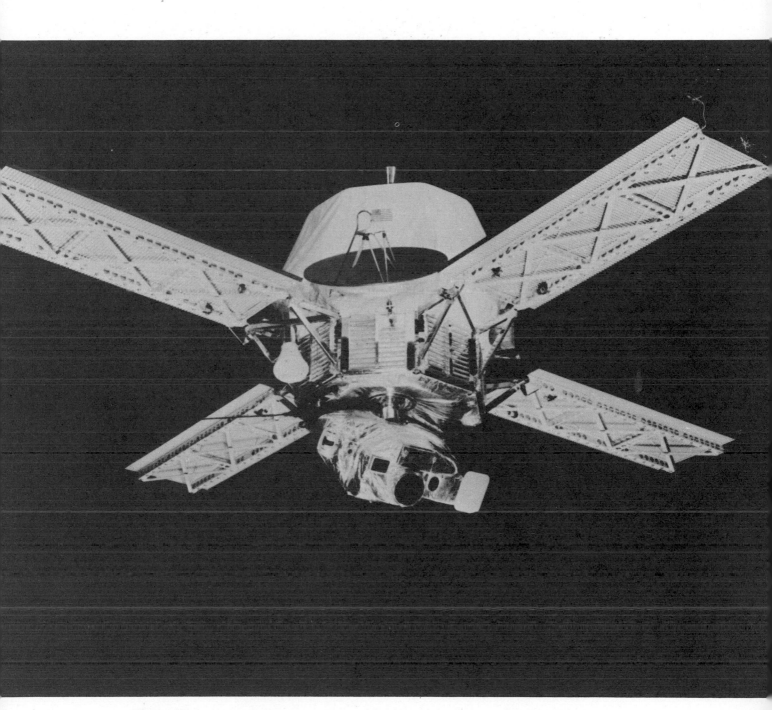

Mariner IX.

right and the critics were wrong. But the scientist was so embittered by the harsh experience that he shut himself off from public scrutiny. Working on his own, he built and flew the world's first liquid-fuel rocket in March of 1926, a 12-foot-tall missile which reached a height of 41 feet above his Aunt Effie's Massachusetts farm, arcing to a point 184 feet from its launch site and attaining a speed of 60 miles per hour. It was a promising start. Subsequent flights were more ambitious, and even inspired flight pioneer Charles Lindgergh to the extent that he convinced the altruistic millionaire Daniel Guggenheim to advance Goddard $50 thousand for further research. With this money, Goddard, his wife, and four aides were able to move to the isolated plains of Roswell, New Mexico, where they built a small rocket range and laboratory. Miraculously, within 15 years, the group had their projectiles screaming at near-supersonic speeds to heights of two miles; over 200 patents were granted to the scientist for his work. Unfortunately, the end to the Goddard story is somewhat downbeat. Just before his death in 1945, the physicist became troubled by the fact that his well-intentioned labors had indirectly helped German scientists build their V-2 rockets, explosive weapons which were used against Great Britain during the Second World War.[1] It was a doubly bitter blow for Goddard, since the United States War Department had repeatedly turned down his pleas to develop rockets for Allied use. However, Goddard would have been crushed even further by American interest in rockets even after the war. The army created a rocket development group headed by such former V-2 scientists as Wernher von Braun, which set up operations in White Sands, New Mexico. The purpose of this operation was primarily strategic, to design antiaircraft and intercontinental ballistic missles with only peripheral attention to strictly scientific programs. By anyone's reckoning, it was a half-hearted effort compared to what the Russians were doing. Soviet soldiers had managed to capture many of the unfired V-2 rockets and hauled them home to Moscow. There, scientists started the ball rolling on a project that was far more bold and portentous than any missile experiments in which the United States was then engaged.

[On one aspect of projectile experimentation, a French scientist at the Eighth International Astronautical Congress in 1957 was moved to

1 In addition to the pilotless, jet-propelled V-1 buzz bombs, and the devastating V-2 rockets—in which the V stood for *vergeltungswaffen*, or *weapons of reprisal*—Hitler's scientists were developing the A-10. This long-range rocket would have been capable of dropping over a ton of high explosives on any target in the world. It's first strike was to have been against New York City.

110

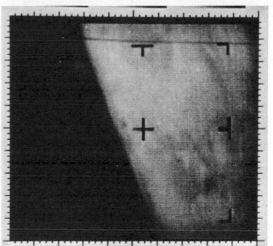

M-IV 01 4 000GC 07/16 20 50 495

This is one of the
first closeup pictures of Mars,
taken at exactly 5:18:33 P.M., PDT,
on July 14, 1965, by Mariner IV.
It shows a bright
region near Trivium Chrontis.

remark, "The Americans talk about it and the Russians do it." He was referring to the launching, on October 6 of that year, of humankind's first artificial satellite, the aluminum-shelled, 184-pound *Sputnik*, a spheroid 22.8 inches in diameter which sent out a "dash" signal three times every two seconds. Both professional and amateur radio operators heard it; telescopists the world-over saw it. Sputnik was a *most* visible piece of propaganda, and everyone was thoroughly impressed. Well, *almost* everyone. United States scientists were alternately impressed and embarrassed: our satellite hardware was still in bits and pieces on workshop shelves. As for political reaction, it was succinct: Vice-President Richard Nixon saw the success of Sputnik as something to remind "both the Communists and the Free World of the increasingly terrifying aspects of modern warfare." Of course, he was really saying that the Russians could just as easily have dropped an atom bomb on the North American continent by simply pressing a button. They had perfected the brute boosters with which the United States was still experimenting. And when Sputnik II was launched on November 3, 1957, carrying the dog Laika into orbit, it was painfully clear just who led the newborn, previously low-gear Space Race. Unfortunately, things got worse for the United States before they got better. On December 6, 1957, a Vanguard rocket was slated to lift the first American satellite into orbit. A recklessly hurried project, the rocket ignited, rose a few feet into the air, then fell back into its own exhaust, buckling and exploding in a cloud of flame and black smoke. The nation was dispirited—but it dared not accept defeat. At the rocket launch site in Cape Canaveral, Florida, scientists erected a second and different vehicle, the Jupiter-C rocket, and sent it aloft on January 31, 1958. The booster's payload was a satellite named *Explorer I*, and the 18-pound scientific package successfully achieved earth orbit. The United States followed this flight with the

111

firing of other satellites, as well as the suborbital journeys of rhesus and spider monkeys and, in the manned *Project Mercury* space capsule, the chimpanzee Ham. Miraculously, after a faltering start, the National Aeronautics and Space Administration seemed to be gaining on the Soviets—which is when Russia put a man in space.

[On April 12, 1961, the Vostok I spacecraft bore the late Major Yuri A. Gagarin on a 108-minute orbit of the earth.[2] In light of this spectacular achievement, the suborbital Mercury flights of America's Lt. Alan B. Shepard, Jr. 23 days later, and Capt. Virgil I. Grissom on July 21, were notably anticlimatic. Even then, the Russians were ready with yet *another* blow to flagging American prestige. On August 6, Major Gherman S. Titov flew an incredible 18 orbits on board Vostok II. As a result of this dazzling piece of international and outer space showmanship, NASA quickly cancelled—or *scrubbed*, as they say in the space industry—four more planned suborbital flights and, on February 20, 1962, sent Lt. Col. John Glenn circling thrice about the earth. However, Russia *again* upstaged the competition by rocketing Valentina V. Tereshkova aloft and making her the first woman in space. Working frantically to keep pace with the Russians, American scientists closed out the Mercury program with a three-orbit flight by M. Scott Carpenter (May 24, 1962), Walter Schirra's six revolution journey (October 3, 1962), and L. Gordon Cooper's 22-orbit ride (May 16–17, 1963), and made ready to unveil the more ambitious two-man capsule *Gemini Program*. But the Russians weren't ready to surrender the race, and took a giant step forward with their Voshkod I (October, 1964) and Voshkod II (March 18, 1965) missions. The first Voshkod was a three-man craft of the sort that was still on NASA's drawing board, while the second craft was a two-man vehicle—launched five days before the first manned Gemini flight—and it allowed Lt. Col. Aleksei Leonov to actually step from his capsule and take a 10-minute walk in space! Thus, at this very one-sided juncture in the Space Race, the ultimate goal of which was to land men on the moon, NASA cried *enough!* This was the last Soviet humiliation they intended to endure. In June of 1965, space officials allowed the late Edward White to space walk for 20 minutes during the Gemini Four flight.[3] Eight more Gemini missions followed, each one more spectacular than the last, with rendezvous' between crafts, dockings with target vehicles, and bril-

2 In 1968, cosmonauts Yuri Gagarin and Vladimir S. Seryogin were killed in a plane crash.
3 On January 27, 1967, a fire swept through the Apollo One command module, killing Virgil I. Grissom, Edward White, and Roger B. Chaffee.

liant orbital maneuvers. After the successful completion of this schedule, the three-man *Apollo* moon program was begun. From this point forward, the Russians lost ground to their Cold War counterparts and never regained the astronautic lead.[4] In 1967–69 there were four three-man Apollo flights—two in lunar orbit—followed by the end of the Space Race when the United States successfully returned two men from the surface of the moon in July of 1969. This historic feat was succeeded by five more American lunar landings over the next two years, after which the Apollo Program was followed by a trio of lengthy stays in earth-orbit onboard the three-man space station Skylab. Most recently, in July of 1975, as a symbolic gesture of what is hoped will be future cooperation in space, the Soviet Union and the United States partook in a joint mission, the Apollo-Soyuz flight, in which three American astronauts and a pair of Russian cosmonauts docked their space vehicles in earth orbit.

While all of this manned voyaging was going on, however, the Russians and the Americans were both busy sending robot probes into space. Beyond the communications and research satellites, there were moon shots. For instance, on October 7, 1959, the Russian craft Luna III beamed back pictures of the never-before-seen far side of the moon. Six American *Ranger* crafts were less fortunate, missing or failing to photograph the moon during their planned 5,850-mile per hour approach and impact with the lunar surface, between 1961 and 1964. Finally, the seventh (July 31, 1964), eighth (February 19, 1965), and ninth (March 24, 1965) Rangers were able to relay photographs to the California-based NASA arm the Jet Propulsion Laboratory.

The idea of using unmanned instrument packages to explore outer space holds an odd place in the hearts of space scientists. If properly designed and constructed, a robot in space can be as effective as an astronaut. A machine like the lunar robot Surveyor or its Martian counterpart Viking, can snap photographs, analyze the soil, sample and dissect the atmosphere, and so forth. They are also less expensive to build than manned vehicles, since they are not laden with living quarters or life-support systems. Conversely, robots in

4 While it is unclear just why the Soviet space effort began to slow down, it is likely that a lack of polished instrumentation was the cause. The Russians had the *power* to put their vehicles into space, but not the technology to build sophisticated space crafts. Indeed, over the next few years, the Russians endured several sharp disappointments in their manned space program, including the deaths of four cosmonauts in faulty returns from earth-orbit missions. Most recently, the Soviets have been conducting a robot space program, in September of 1976 landing a probe on the moon, having it drill for samples, and then successfully returning it to earth.

space just don't have that spark of romance which we found in the fiction of outer space. This is all-important in firing public interest and, thus, Congressional support for the space program. While people can easily identify with a heroic astronaut, drumming up emotion for a transistorized, space-voyaging laboratory is considerably more difficult. But the question is becoming more and more lopsided in favor of the unmanned explorers. Indeed, in these early years, when we are talking about planetary exploration rather than colonization, the argument for machines is almost airtight. Let's look, for example, at what robots have told us about Mars.

[Five Mariners, six Russian Mars crafts, and two Vikings have thus far been launched at the Red Planet. Only eight of these instrument packages successfully completed their tasks, but, together, they have told us more about Mars than all the telescopic studies since the time of Galileo. It began in 1964 when both the Space Science Board of the National Academy of Sciences, and the National Research Council advised NASA that, after a manned landing on the moon, an unmanned search for life on Mars should be the agency's foremost objective. Venus was not to be a prime target since both Russian and American probes launched in 1962 had shown the planet to be utterly inhospitable to life, with temperatures reaching 900 degrees F., and a surface pressure 90 times that of the earth. Contrarily, exploration of the more temperate Mars would be a valuable biological program, and it was something that would presumably capture the taxpayers' fancy—particularly if canals or tharks were discovered on Mars. And public relations are an important aspect of selling pure science to the generally disinterested populace. Unfortunately, the assault on Mars had gotten off to a dubious start, to say the least. The Russians' Mars I, which was launched on November 2, 1962, was to have passed within 115,800 miles of Mars, sending back photos and data: its radio died midway through the trip, and it was a frustratingly silent craft that glided by the Red Planet. Mariner III, launched for Mars, was equally ill-fated. The probe blazed into space on November 5, 1964, but the fiberglass cover which protects the Mariner from the stress of lift off, failed to disengage, rendering the craft useless.[5] Fortunately, Mariner IV was far more successful. The stubborn shroud that

5 Mariner I, bound for Venus, was ordered to destruct over the Atlantic Ocean when the booster went awry. Mariner II, a nonphotographic probe, flew past the second planet, as did the Mariner V craft. The picture-taking Mariner X not only made it to Venus and returned incredible photographs, but it went on to Mercury and beamed back astonishing shots of that heavily cratered world.

114

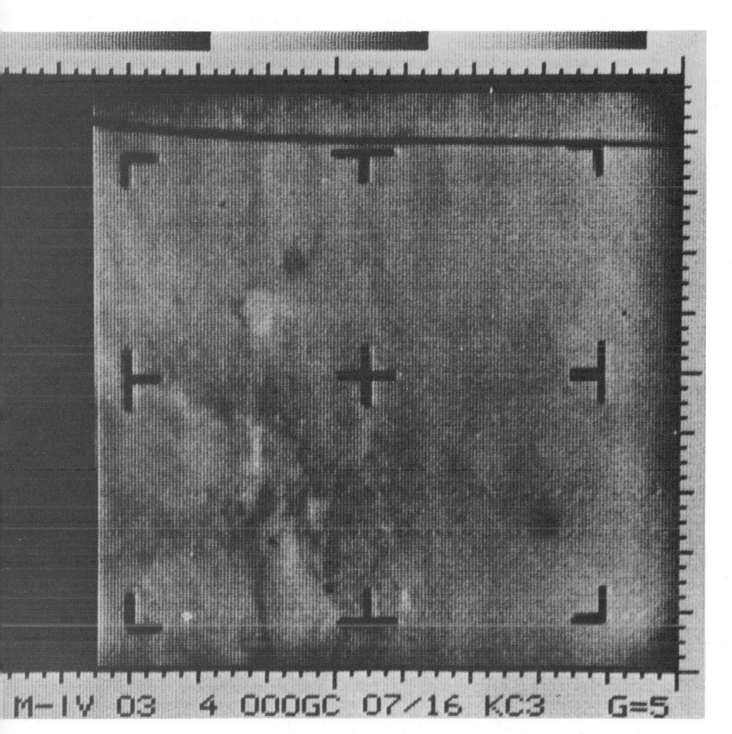

M-IV 03 4 000GC 07/16 KC3 G=5

Another Mariner IV photograph showing the region southeast of Trivium Chrontis.

had set its predecessor afoul was hurriedly replaced by a more cooperative magnesium alloy model, and the robot was sent aloft on an Atlas-Agena booster at 9:22 A.M., EST, on November 28, 1964. And, although the space lab was laden with all manner of scientific instruments, it was a television camera that was to have the greatest impact, snapping 22 photographs on July 14, 1965, when the planet and probe intersected, it showed us a Mars unlike anything of which scientists had ever conceived. Let's recreate the setting just prior to that transmission from Mars.

[Mariner was near the end of its 228-day journey, its orientation in space having been maintained by Canopus, a star in the constellation of Carina (the Keel, a portion of the constellation Argo, the mythological vessel of Jason), over 200 light years away.[6] Not that Mariner had welcomed Canopus with open sensors! Six days after launch, scientists at the Jet Propulsion Laboratory, the interpretation and control center of NASA's unmanned flights, found that their Mariner was on a course which would bring it to within an unsatisfactory 151,000 miles of the Red Planet. Thus, a correction was necessary, one which would cause the craft to shift its trajectory using Canopus as the point of reference. Unfortunately, before the maneuver was sent to Mariner, the craft "blinked" and lost the star. As per its programming, the probe began to roll about in search of the beacon. And, showing just how fickle a comparatively unsophisticated space craft like Mariner can be, it fixed on seven different stars in succession, mistaking them all for Canopus. Each time, the JPL had to signal that the star in its eye was *not* the Yellow Giant. Obediently, Mariner would then roll on in search of the guiding light. Finally, after nearly a full day of interplanetary coaxing, Mariner was back on the track, and a 22-second "burn" of its engines put the craft on a course that would bring it to within 6,118 miles of Mars. Thus, after traveling 325 million miles through space, Mariner's journey was at an end.[7] Gathered about the photographic instruments in a crowded, anxious Jet Propulsion Laboratory on the morning of July 15, 1965, were physicist Robert Leighton, geophysicist Robert Sharp, geologist Bruce Murray, and JPL

6 A light year is the distance traveled by light, at a speed of 186,300 miles per second, in a year; approximately six trillion (6,000,000,000,000) miles.
7 Mariner IV traveled well over a quarter of a billion miles to Mars because it did not simply fly from point to point. In order to rendezvous with the planet, it went into solar orbit, launched along such a course that it would intercept the path of the Red Planet 228 days later.

116

technicians Richard Sloan and J. D. Allen. They were waiting for Mariner to begin transmitting the 240,000 pieces of information which translated into 40,000 binary digits representing 64 possible shades of black, white, and grey, zero being the brightest and 63 being deep black. These would be rebuilt on earth as a single photograph, an image composed of 200 tone dots per line, and 200 lines per picture. The conversion would be worked by a computer transposing the Mariner's bits of information onto magnetic tape that, when scanned with a beam of light, changed the number to its corresponding visual shade. However, since Mariner's 10-watt radio required eight and one-half hours to send each photograph to earth, at a rate of eight and one-third bits of information per second, the wait for this monumental photograph would be a long one! The signal to position its lens had been sent to Mariner at 10:10 in the morning on July 14, an order which took 12 minutes to cross interplanetary space and reach the craft. Then, at 5:20 P.M., when an on-board electric eye first saw the edge of the planet, the ship's preprogrammed computer alerted the camera to begin taking the 22 pictures. Twenty-five minutes later, the photographic survey had been completed and stored on tape. Or had it? At once, technicians feared that things had not gone as planned. The tape, which recorded and coded the photographs inside Mariner, was supposed to run for one-fifth of a second when the camera was in operation, stop, and begin again a minute later when the Mariner camera was due to take the next picture. Unfortunately, information relayed from the space craft to the JPL suggested that the tape had *not* stopped between exposures. If this were so, then not only would the tape have run out long before all the photographs had been taken, but there was the chance that those which *had* been recorded were obliterated by the spools running on and on. To at least safeguard against this latter problem, JPL scientists ordered the tape system to shut down, if it hadn't as per its preprogramming, until transmission to earth was scheduled to begin the following day. At that time they would learn if Mariner had been able to gather its full complement of pictures.

[On the morning of July 15, the pictures began their long voyage home. That night, Prof. Leighton projected the first photo for reporters and scientists: it was not the breathtaking shot everyone had been hoping for. The very dim shot showed a 200-mile stretch of the edge of Mars known as Elysium. No one was particularly impressed since

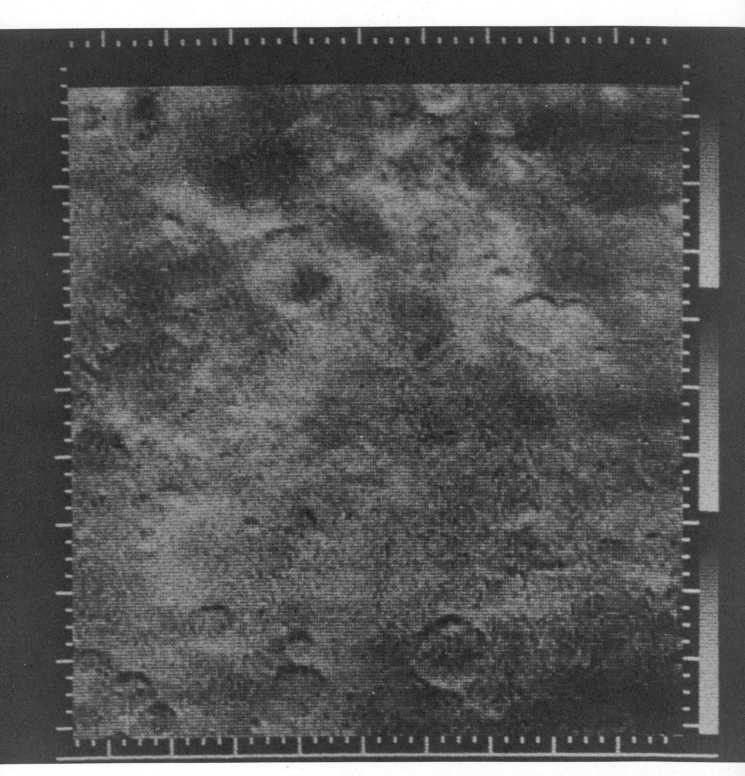

Mariner IV took this 180-mile-by-180-mile view of southeastern Zephyria, near Mare Sirenum, nearly eight minutes after its brief photographic rendezvous of Mars began.

all that could be seen were a few dark smudges on an equally unclear plain. Then, Leighton confidently presented a refined version of that first photograph, one which had been reinterpreted by a computer and its inherent detail enhanced. Now the reaction was quite different, and there was a spontaneous wave of applause. However, the dimly seen rills and cracks of the first, then second and third photographs, while they suggested a rougher terrain than scientists had expected, were nothing compared to what appeared over the next week and a half. As it turned out, 21 pictures and a portion of the 22nd had been recorded, so fears about a faulty tape system had been unwarranted. But what these pictures showed made everyone forget that there had ever *been* a problem. In fact, looking at the Mariner photographs, many scientists were no longer sure that they knew anything at all about the solar system! The 1 percent of the Martian surface scanned by Mariner was covered with jagged mountains, valleys, and *craters*, not unlike those of the moon! Craters, ranging from three to 75 miles across and some of them thousands of feet deep. The pictures were rushed to Washington, D.C., where they were viewed by President Johnson and members of the Congressional Space Committee. Beyond their immediate scientific impact, the photographs meant something very significant to the President. He put his reaction into perspective by noting that he belonged to "the generation that Orson Welles scared out of its wits." He went on to say, with considerable reverence, that "it may just be that life as we know it, with its humanity, is more unique than many have thought, and we must remember this. I believe it is very clear that in this day, when we are reaching out among the stars, the earth's billions will not set their compass by dogmas and doctrine which reject peace and embrace force and rely upon aggression and terror for fulfillment."

[Apart from its remarkable photographic accomplishments, Mariner IV made other contributions to our knowledge about Mars before passing the planet and becoming a tiny satellite of the sun. The probe determined that Mars has no magnetic field, a force generated on earth by the affect of rotational energy on the molten iron core. Since Mars rotates in roughly the same fashion as the earth, this discovery implied that Mars does not have a solute iron center. Nor, as a result, could there be a radioactive band such as the terrestrial Van Allen Belts, since these are caused by protons and electrons being trapped by the earth's magnetic field. Other experiments showed the Martian atmosphere to be much thinner than scientists had anticipated, as little as 1 or 2 percent that of the earth. However, Mariner did find an ionosphere, which means that Martian settlers will one day be able to bounce radio waves against this electrically charged atmospheric layer to distant points on the planet, just as we are able to do on earth. Of course, the question of colonizing Mars seemed terribly moot after the revelations of Mariner IV, and even more so after the voyages of Mariners VI, VII, and especially IX. Not only did the Red Planet appear hopelessly *dead*, but it was increasingly apparent that nothing from earth could hope to survive on the surface of Mars without the protection of a self-contained vehicle or space suit.

[The Mariner VI craft, weighing 910 pounds, or 335 pounds more than Mariner IV, was launched from Cape Canaveral[8] on February 24, 1969, and passed Mars at a remarkably precise distance of 2,131 miles on July 31, 1969—12 days after Apollo 11 had landed on the moon. A duplicate space craft, Mariner VII, sent Mars-ward on March 27, 1969, repeated this 2,131 approach on August 5 of that year. Between the two explorers, they took 202 photographs, mapping 9 percent of the Martian surface. They uncovered many nonterrestrial *and* nonlunar markings—in other words, generically *Martian* features such as expanses of inexplicably chaotic ridges and tortuous crevices, butted by smooth terrain with soft, eroded craters. Pictures of the south polar cap found what appeared to be a thin layer of snow, which was thought to be frozen carbon dioxide. However, enlightening though these Mariners were, they did not provide a continued flow of information about the fourth planet. A brief crossing of paths with Mars, and then it was off into solar orbit for these game little vessels. Thus, NASA decided that their next Mars crafts would simply have to go into planetary orbit. This was to be the mission of Mariner VIII and Mariner IX, but almost at once there were problems. Scientists like Bruce Murray at the California Institute of Technology, the university adjunct to the Jet Propulsion Laboratory, were interested in the geophysical history of Mars, and requested that one of the two orbiters be inserted into a polar orbit. They felt that the little-explored ice

8 Cape Canaveral, Florida, was renamed Cape Kennedy in honor of the slain President. However, many of the local citizens never stopped using the original designation, and it was recently reinstated.

This magnificent globe of Mars was constructed using 1,500 photographs taken by Mariner IX in 1971 and 1972. It is the first mosaic-photograph globe of any body in the solar system.

caps would tell us much about the geology, seismology, volcanism, and past or present oceanography of Mars. Conversely, Carl Sagan and the space scientists from Cornell University in Ithaca, New York, were more interested in the darkening maria and the equatorial regions of the planet. They wanted explanations for the phenomena that were visible through earth-based observatories. These requests in themselves presented no problem: one Mariner would go to the Martian equator, the other to one of the two poles. That's when the 2,200 pound Mariner VIII, launched on May 8, 1971, rose majestically from its launch pad and went roaring into the Atlantic Ocean. If anyone had been thinking along those lines, perhaps the sunken probe might have told us a great deal about our own planet's sea beds. Unfortunately, no one could see past the cloud of depression that shrouded the JPL, for *now* there was trouble! We had the 2,150 pound Mariner IX ready to fly, but the question was into which Martian orbit would it be placed? Despite fevered pleas from both camps, NASA mission directors had only one logical choice to make: to put the vehicle in a 65-degree path around Mars, half way between the potar and equatorial regions. Scientists on both sides of the dispute grumbled, but it was a situation with which they were forced to live. The craft was rocketed toward Mars on May 30, 1971, and became a satellite of the planet on November 13 of that same year. And, despite its compromise orbit, Mariner IX became the JPL's invaluable eye in space for nearly a year. Incidentally, this lengthy period of activity came as a shock to the people who built the probe. The engineers said that once it had reached the Red Planet, Mariner's optimum life expectancy would be 90 days! Thus, Mariner IX was eventually able to snap 4,000 pictures of Mars, covering the entire planet through orbital changes, and making all the scientists happy. Of course, as per an unfortunate NASA "tradition," the gathering of this data was not without its tense moments! Only this time, they had nothing to do with technical or human error.

[When the robot satellite arrived at Mars, it found the Martian surface obscured by a planet-wide dust storm. *Not a single feature was visible*, nor would it be for another *two months!* NASA's face was redder than Barsoom at sunset, since they had promised to take our breath away with the sights seen by the Mariner. Fortunately, Carl Sagan had an ego-saving proposal which proved to be a stroke of genius. Over a year before the launch of Mariner IX, Sagan had suggested that it be allowed to take a brief photographic look at Phobos and Deimos, the moons which appeared as pinpoints of light in even our most powerful terrestrial telescopes. Beyond what this sideward glance might tell us about Mars and its tiny space companions, it might also save NASA a proposed $200 million robot-trip to the Asteroid Belt, planned for the 1980s, if, as some scientists speculated, the satellites were actually planetoids captured by Mars' gravity. But space agency officials were opposed to the idea. The detailed mission plan had already been drawn up, and changing it would be an inconvenience. But inconvenient or not, when Mariner arrived at Mars and all NASA had to show the press for its efforts were a host of near-blank photographs, they hastened to turn disaster into technological triumph. As if it were a grand inspiration, they ordered their explorer to start taking pictures of the Martian moons. The aligning of Mariner's television eye with its chosen target, Phobos, required two weeks and, on its 31st orbit, the space probe radioed a picture of the moon back to earth. The image showed a hazy blob of light and nothing more; computer enhancement did little to improve the image. A second photograph, taken two days later from 4,000 miles away, was of a better quality but still lacked detail. Not surprisingly, three people—Sagan, his former student Dr. Joseph Veverka, and a JPL technician—worked late that night to reprocess this new transmission. The men watched as the picture became clear— and the moon showed itself to be, as Sagan described it in his book *The Cosmic Connection*, "[like] a diseased potato . . . heavily cratered." It's a perfect description of this lump of rock which Mariner accurately measured at 13.6 by 11.2 miles. The more regularly shaped Deimos was subsequently catalogued at 8.5 by 7.9 miles in size. However, as Sagan had predicted, in telling us a great deal about the two moons, Mariner also enhanced what we knew about Mars. For instance, because the moons have neither air or water, the oldest craters on their surface have been undisturbed for billions of years. By comparing these to similar craters of Mars, we get a general picture of the way in which the Martian environment has affected and eroded its own surface formations over the years. These moon markings also provide us with pieces to the puzzle of the creation of the solar system, their battered surfaces lending strong support to the built-by-collision theory of the planets as described in chapter two. As for Phobos and Deimos being captured asteroids, the question is still unanswered. To be sure, we don't expect asteroids to be very different from the small Martian moons. However, the Mariner IX photographs *did* solve one puzzle about such bodies. It was long thought that asteroids struck by meteoroids might have pocked rather than cratered surfaces, the difference being that craters have raised rims formed by the impact-ejected debris. Because of their negligible gravitational pull, it was believed that asteroids would be unable to counteract this force and hold the shattered pieces about the pit. As it turns out, although we don't yet understand *why*, fully blown craters were found on the two moons. And they have certainly given the geophysicists something to think about! More on the moons when Viking makes some awesome observations in our next chapter.

[Early in 1972, the weather began to improve on Mars proper, as the dust storms subsided and the planet's surface became visible.[9] Thus, over the next 10 months, Mariner IX was able to send 7,232 photographs of Mars to the JPL, pictures of amazing clarity and immeasurable importance. They showed *canali*, but not of the sort seen by Lowell: these were riverlike threads, winding fluidly across the Martian terrain and suggesting that they had been carved by water. This meant that at some time in the past, great seas and streams may very well have covered sections of the planet. However, if, at one time, there were bodies of water on Mars, Dr. Harold Masursky, of the United States Geological Survey, believes that the episodes were brief, lasting maybe a few thousand years. He places these periods at 2 billion, 500 million years, 1 billion years, and 100 million years in Mars' past. But water also implies precipitation, which Dr. Masursky describes as follows: "A cloud burst would occur over one spot and the rains would fall heavily, eroding the hell out of that particular spot. It was a sporadic process, sporadic in time and place, and that would account for the fact that we see both water-cut channels and craters that somehow avoided being eroded away by that water." This, of course, is speculation. More concrete evidence pointing toward water on a younger Mars was found in the nature of

9 The dust storm was not a total debacle for the Mariner scientists. Their probe told them that every Martian year, at the planet's closest approach to the sun, a dust storm begins where the regions Hellespontus and Noachis meet. The winds grow in strength and move in a westerly direction, eventually blanketing the entire planet.

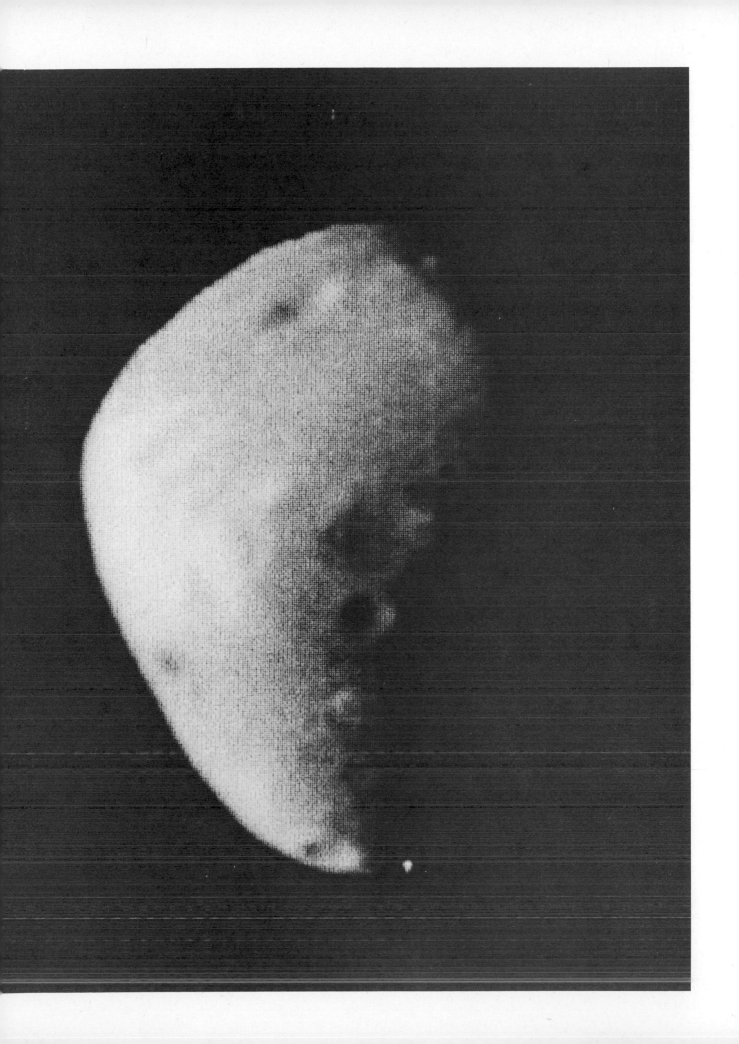

regions surrounding the polar caps. Their deteriorated condition was strikingly terrestrial, meaning that they may once have been subject to some form of glaciation. But where, then, did all of this water go? Some scientists felt that most of it had fled into space as the Martian atmosphere dissipated (assuming that the air blankets were once thicker than they are at present); others maintained that it is hidden beneath a layer of Martian dust. Still others, with conflicting interpretations of photographs and spectroscopy in hand, insisted that there *is* liquid or vaporous water on Mars, and that it is simply a matter of finding it. While Viking would shock everyone with some solid solutions to the controversy in September of 1976, that water once coursed about the planet seemed an inarguable fact as of January 1972.

[Another seat of considerable disagreement among scientists concerned the composition of the planet. No one knows exactly what the rocks and dust of Mars are made of, although Mariner findings suggest a few answers as well as a history of Mars. Current evidence indicates that Mars was built in more or less the same fashion as the earth, by 100,000 years of colliding space fragments. As a result of this accretion, Mars was heated by its cumulative gravitational energy. Thus, while the molten outer layers of the planet cooled to form a 33-mile-deep crust, rich in silicon and aluminum, radioactive decay caused the elements beneath it to separate into gases. These escaped to the surface, such matter as iron and iron sulfide became a dense core, and ferromagnesian silicates formed a mantle. Here, however, is where Mars and the earth pursued different evolutionary courses, and it is a period about which scientists are unclear. Gases in the Martian atmosphere obviously underwent many of the same changes they did on earth, with the exception that varying proportions of some gases along with the inability of Mars' gravity to hold them all led to a significant difference. If there *were* seas on Mars, and if they did spawn life, then they were both undoubtedly destroyed by the fact that Mars was unable to hold hydrogen—the H of H_2O (water)—a fact revealed by the ultraviolet spectrometer aboard Mariner. The instrument tracked hydrogen even as it was scattering into space, and at such a rate that, if it were constant, Mars would at one time have had enough water to cover the planet to a depth of several miles. This water break up would also have freed the oxygen in water so that it could oxidize matter in the Martian crust. Conceivably, it is this millennia-long process that has given the Martian surface its rusty color. Another puzzle involves the volcanoes discovered by Mariner IX. For some reason, the handful of huge, possibly active volcanoes lie in the northern hemisphere. The craters first revealed by Mariner IV all occur in the south. It is a puzzling distribution, and one which may not be understood for years to come. But we do have some facts and hypotheses about these features themselves.

[The largest volcano on Mars is Olympus Mons, a mountain which had in fact been seen from earth and named Nix Olympica. However, not until Mariner did scientists have any idea as to the mountain's true nature! Mariner's revolutionary readings told us that, for one thing, Olympus Mons towers 15 miles above the surrounding plain—earth's tallest mountain, Everest, is just shy of six miles in height, and it's not even a volcano![10] The probe also detected what appear to be ribbons of gas crawling from the volcano, which indicate that it may still be active. Nearby, along the Tharsis Ridge, is another trio of volcanoes, all of which, like Olympus Mons, are acutely angled slopes built by years of rolling and hardening lava flows. A second type of volcano can also be found in Tharsis, peaks with more sharply inclined sides than the so-called shield volcanoes, leading scientists to believe that they are younger and have not yet had the outpourings of molten rock which formed Olympus Mons and its companions. A third and final volcano type found on Mars is centralized in the Elysium and Tharsis regions both, and is little more than a barely raised crater. Their origins are a puzzlement, and are likely to remain so for some time.

[In themselves, as seen through the eye of the orbiting Mariner, the volcanoes tell us very little about Mars other than the fact that it was and may still be a hot, active planet.[11] However, the adjacent plains, which are obviously beds of hardened volcanic issue, are informative. Many of these areas overlap crater-riddled expanses,

10 Scientists believe that the volcanoes on Mars are larger than those on the earth because of the Red Planet's unique tectonism. Tectonics pertain to the structure of a planet's crust. In the case of the earth, the crust is made up of plates which move one against the other, causing changes in the earth's geography. Because of this motion, the layers beneath these plates, which supply the volcanic building blocks, don't have the chance to vent material through one outlet only. On Mars, however, there is no indication of such horizontal movement. Thus, the eruptions occur in isolated areas, creating such landmarks as the Olympus Mons.

11 The question of current volcanic activity on Mars is still up in the air, as it were. Mariner IX's instruments were not sensitive enough to find lava flows, nor could they have detected a comparatively small "hot spot" on the planet's surface.

while in other instances—primarily in the south—the planet remains badly scarred without relief. This indicates that the planet was widely bombarded before its period of severe volcanism, although said volcanoes obviously left the south largely untouched. Since Viking, our current Martian-lander probe, does not contain a geological laboratory, it will take more intricate spacecrafts to accurately determine the length of this eruptive activity, and why it was centralized in the north. Only then will we be in a better position to understand the physics and duration of the collisional process that formed both Mars and the other planets.

[Apart from volcanoes, Mariner IX presented scientists with the perplexing discovery of Martian troughs. These huge gashes in the planet's surface would, under comparable terrestrial conditions, be referred to as canyons. However, not all of them are open at both or even one end, nor are they in any way linked with what might once have been a river or sea. Most of these troughs are situated above the equator, and many of them are up to 100 miles long by 150 miles wide. Geologists have no idea what caused these great tears in the crusts, although the most popular theory is that they were created by the shifting of land situated between two faults. Perhaps we shall learn more about these as we study other rifts in the planetary terrain, those that more closely resemble valleys on earth. With the sinuous outlines of what once must have been flowing rivers winding to, from, and throughout the depressions, their origin is no doubt water-related. Long-vanished layers of ice may also have contributed to the formation of these striking features.

[Thus far, we have measured the Martian topography against comparable earth features. However, one geological specimen is decidedly more moonlike than terrestrial, the craters of Mars and their big brothers the basins. The craters range in size from 10 to 20 times greater than the interplanetary fragment that formed them. One reason the Mars was particularly susceptible to such bombardment was its proximity to Jupiter, whose immense gravity forced a dramatic movement among space rock and debris in the Asteroid Belt. Too, Mars' atmosphere does not disintegrate charging meteroids as effectively as the heavier air blanket surrounding the earth.

[Craters smaller than 7 miles wide usually have a raised rim and are surrounded by matter thrown with the impact. The larger craters, the basins, are flat, walled expanses ranging from 150 to 1,000 miles

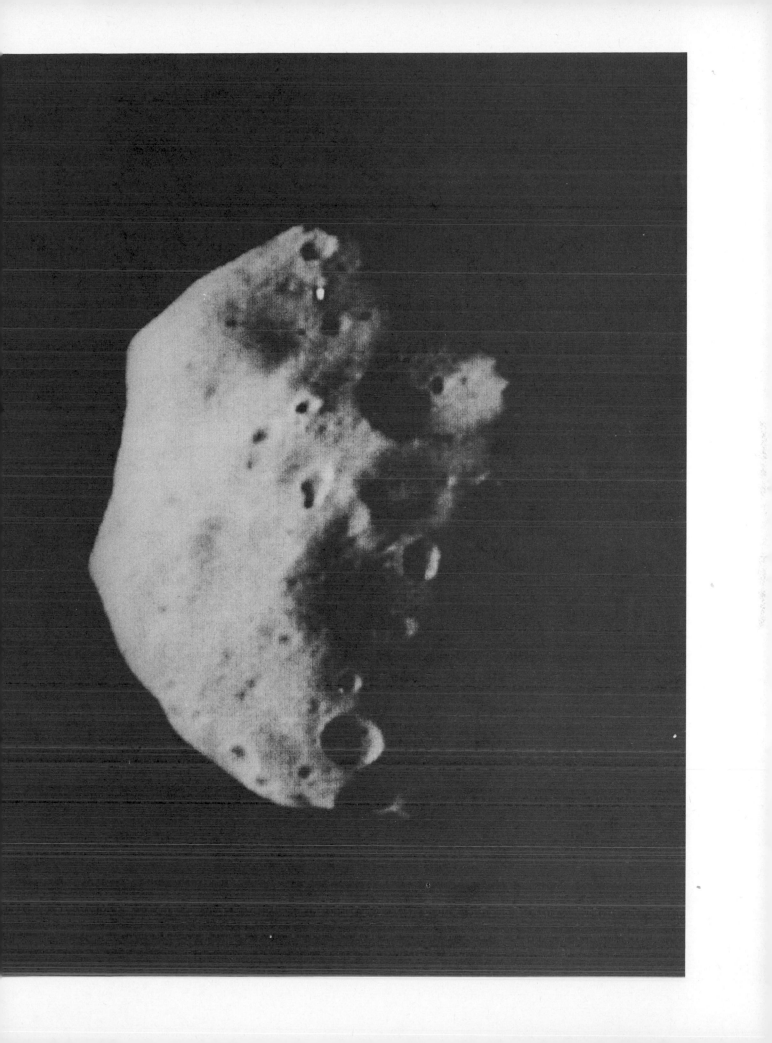

in diameter and are believed to have been the result of more cataclysmic contact early in Mars' history. The largest such basin, Hellas, is 1,320 miles across and 2.6 miles deep.

[The last of the major Mariner IX contributions to our knowledge of Mars were more detailed analyses of the atmosphere and weather than we had been able to conduct from the earth. Such tools as the infrared radiometer and the ultraviolet spectroscope read the atmosphere and found it to be composed of up to 90 percent carbon dioxide. They also sensed the presence of clouds, many of which contained water vapor. As for specific surface pressures, in the lowest chasms, it was discovered to be nearly one percent that of the earth, and .3 percent at higher elevations. A human being, placed unprotected in such an environment, would find the pressure insufficient to allow for breathing—even if there were enough oxygen to sustain terrestrial life—and that, in any case, his or her blood would boil in a matter of seconds. Finally—and perhaps most dramatically—those dust storms which had been whipping about Mars early in the Mariner IX mission, were clocked at up to 320 m.p.h. Due to the lesser atmospheric pressure and gravity, this is roughly equivalent to a force of seventy m.p.h. on earth—a considerable wind indeed! An astronaut stranded in such an holocaust would find the fiercely tossed particles of Martian dust surprisingly destructive to both spaceship and equipment.

[Mars had certainly been a less perplexing planet some 60 years before this dynamic Martian Mariner. Canals, intelligent beings, awesome fields of vegetation: it was all so clear, something the wonder of which even the layman could grasp. Now, Mars belongs to the scientist. However, it is only an interim possession. The Red Planet will be returned to the taxpayer, for his use, as we follow up the knowledge that Mariner has given us of the real Mars; learn more about the planet's climate, geology, and life forms, if any. Then we will "terraform" the world—make it hospitable for human beings. As we will soon see, the earth is a terribly finite place. Unless we colonize new worlds, there will come a time when we as well as our native planet shall cease to exist.

[Apollo 11 was the first such giant leap for humankind; Mariner was a smaller step, but no less important. Now comes Project Viking. And, like its namesake, the robot program will be the first to truly liberate our race from the small lunar-terrestrial point of reference to which we have so long been bound.

6

STOP THE WORLD—
I WANT TO GET ON

Since the end of the moon-landing race with the Soviet Union, NASA has led a proverbial hand-to-booster existence, as decreasing annual budgets have forced them to slash programs and personnel which, in turn, has radically curtailed our exploration of space. Ironically, Viking was born from one such cut. When a lack of Congressional funding killed the ambitious Voyager program that was to land automated scientific laboratories on Mars beginning in 1973, the more modest Project Viking replaced it.

[Before we examine this $1 billion Mars program, it behooves us to answer the inevitable question, "Why spend all that money on space when we have so many problems here on earth?" We will confront this argument—and then dismiss it as ultimately frivolous and uninformed. The basic defense of the space program lies in the simple fact that the earth is a finite world. Within a few decades, our coal and oil resources will have been utterly depleted; we will have overcrowded our planet with people; and eventually, our sun will die, and the solar system along with it. Thus, tying our race to earth is analogous to committing suicide. And, while it can be argued that these catastrophes will occur beyond our own lifetimes, this is a singularly irresponsible attitude to take. If nature has given all animals, including human kind, one clear-cut duty, it is to preserve their species. The space program promises us survival by opening other

131

worlds and other star systems to colonization. But let's come down a
notch from Armageddon. To date, what has the space program given
the ordinary earth citizen? Well, beyond those fancy digital cal-
culators with which everyone loves to tinker—they were designed
with the technology we gained by transistorizing instruments for
space vehicles—there are awesome advances in health care. The
20,000 babies who would normally succumb each year to respiratory
disease in which the lungs lose their ability to oxygenate the blood,
are helped by circulatory aides developed to give the astronauts'
circulation a boost in the weightless conditions of Skylab. Also, the
precise Skylab medical checklist is now the standard manual of
ambulance care. And how does one measure the worth of biological
isolation garments which allow immune-deficient children to leave
treatment centers and go out into the world? Or a compact "black bag"
for doctors, weighing 30 pounds and containing, in addition to the
standard medical tools, an electronic vital-signs monitor, electro-
cardiogram and electroencephalogram devices, and equipment for
minor surgery, which was created from compact tools used on space
missions? One can be certain that the tens of thousands of people
whose lives have been saved thanks to these representative developments will not ask why we bother with space. As for returns on our tax
dollars in more widely affective terms, space technology has given us
new, more versatile, and less expensive building materials such as
tough, noncorrosive paints, and wire that is safer, thinner, and
less expensive to manufacture than conventional coils. Protecting the
buildings in which these materials are used are firemen who now do
battle using lightweight air tanks developed for lunar excursions,
thus increasing their maneuverability inside a blaze. Firemen are
also being equipped with efficiency-building shortrange radios designed for weather-balloon communication. And what *about* the
weather? Advance warning from our earth-orbit satellites of mounting
storm systems and hurricanes has saved hundreds of thousands of
lives; these same eyes in space have charted the migratory patterns of
fish, allowing underfed peoples to spearhead successful catches that
were previously hunt-and-peck endeavors. The fact that we can even
contact citizens of other lands on a massive scale is the result of
hundreds of communications satellites we have put into space. Then
there are studless winter tires built after the wheels on our lunar
rover, tougher brake linings of the same material as Viking's

132

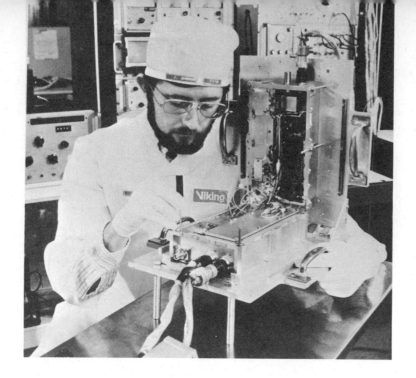

Above: RCA technician Jim Johnson tests the radio that will be humankind's first broadcasting station on Mars. The weight of the unit is ten pounds. *Below:* NASA technicians at the Jet Propulsion Laboratory work on a test model of the orbiter.

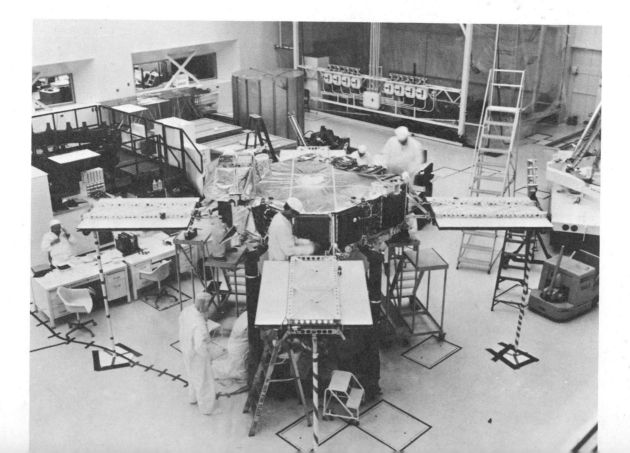

parachute lines, smoother roadways, more effective pollution control and sewage treatment, and nonspoilable space food sent *by mail* to elderly people who cannot leave their homes. These are only a *few* of the immediate wonders made possible by the space program. When we consider the *trillions* of dollars spent on defense and the matching sums squandered on welfare and unemployment—think of all the jobs the space effort has created!—then the worth of the endeavor becomes inarguably clear. Saving our race by becoming a spacefaring people is only the longest range of the space goals. Thus, the program that began with Sputnik, the spreading of our collective space wings, continues, now, with Viking. At the end of this chapter, we will examine the projects that go beyond the Mars lander. In the meantime, inconsiderate of these crucial implications of humankind's thrust into space, and the nobility of both our immediate and long-range plans, there is a specific scientific goal to each craft we launch into space. And Viking's most critical assignment was to learn if there is life on the Red Planet. Our first order of business, then, is to understand just what kind of vehicle can, among its many diverse abilities, look for living things on a world that is tens of millions of miles from our own.

[Discounting the Viking launch vehicle, which consists of a Titan III-E booster and a Centaur D-1T upper stage, the spacecraft is a Martian orbiter weighing 5,125 pounds, 3,097 pounds of which is propellant, and a landing vehicle of whose 2,464-pound weight 307 pounds is fuel. The two crafts fly to Mars as one unit, although each is a distinctly different craft. The job of the orbiter is to photograph Mars while the two explorers are joined in Martian orbit, using its twin cameras to scan potential landing sites, and relay these pictures to the Jet Propulsion Laboratory. It will also look for water vapor using a Mars Atmospheric Water Detector System—a concentration of water might indicate biology, and thus represent a favorable landing site—and will pick out "hot spots" using an Infrared Thermal Mapper, since life, as a rule, seeks out warm environments. After separation, the orbiter performs other duties that we will analyze in a moment. The lander, following a successful touchdown, takes pictures of the Martian surface with its two cameras, searches for indigenous life forms, and performs experiments which we will discuss in greater detail later in the chapter.

[The components of the Viking scientific package were manufactured by different firms across the United States. While the manage-

Another view of the assembly
of the orbiter model.
The two large fuel tanks hold 3,100
pounds of propellant between them.

ment of Viking originates at NASA headquarters in Washington, D.C., the bulk of the Viking decision-making and construction coordination is handled by the 250 scientists and technicians at NASA's Langley Research Center in Hampton, Virginia. Langley oversees the lander system design and construction, which occurs at Martin Marietta Aerospace's Denver Division, the creation of the orbiter at the JPL, the building of the Titan III-E at Martin Marietta Aerospace, and the assembly of the Centaur at the General Dynamics Corporation. When these various elements are completed, the companies shipped them for final packaging to the Assembly and Encapsulation Facility (One) at the Kennedy Space Center in Florida. There, the orbiter and lander are mated, then baked at 236 degrees F. for 40 hours to kill clinging terrestrial germs and prevent the possible contamination of Mars. Such an infestation might be misinterpreted as natural Martian biology. After sterilization, the vehicles are nestled within a protective shroud and placed atop the launch vehicle. The selection of a rocket to lift Viking skyward posed something of a problem for flight planners. The Atlas-Centaur which had hoisted the 2,150 pound Mariner IX aloft would not have the muscle to raise the four-ton Viking. On the other hand, the space effort's real workhorse, the Saturn V lunar rocket, was twice the necessary size. Thus, the relatively new Titan/Centaur hybrid was selected, 159 feet tall and built around a two-stage Titan II core, with a pair of large zeroth-stage rockets strapped to either side of the booster.[1] The Centaur stage has two engines which can be shut off and started again from earth. The entire vehicle is launched by the two side rockets, generating 4,893,230 newtons (1,100,000 pounds) of thrust at lift off. Two minutes later, they are jettisoned over the Atlantic. However, just before this happens, the Titan core ignites. The first stage burns for two and one half minutes; the second stage for three and one half minutes. Its job completed, the Titan then falls away and the Centaur takes over. Its two rockets generate 133,452 newtons (30 thousand pounds) of thrust and lift the Viking compartment into a 90-nautical-mile-high parking orbit around the earth,[2] and waits anywhere from six to 30 minutes for the exact

1 Zeroth means, quite simply, that the rocket bears only itself aloft, and no subsequent stages.
2 For fuel buffs, the propellants used in the two side-saddle rockets are powdered aluminum and ammonium perchlorate; the Titan burns hydrazine, unsymmetrical dimethylhydrazine, and nitrogen tetroxide oxidizer; the Centaur flies with hydrogen and oxygen.

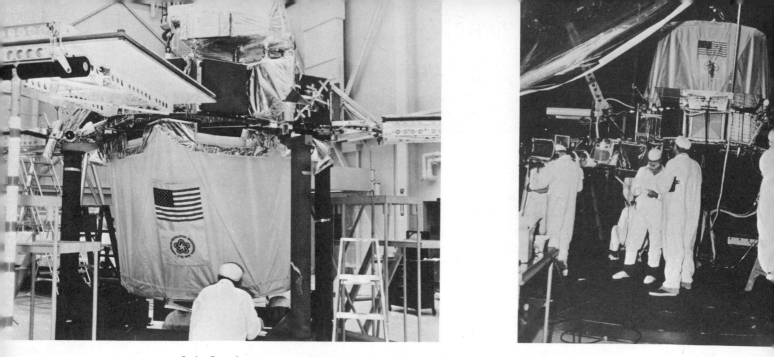

Left: A technician puts the finishing touches on the orbiter model. *Right:* The orbiter model is prepared for thermalvacuum testing in the 25-foot diameter space simulator at the Jet Propulsion Laboratory.

moment when a burn will send Viking soaring along a heliocentric course for the eventual interception of Mars.[3] Then, using what is left of its propellants, the Centaur separates from Viking and blasts into a different orbit entirely. This is done so that the rocket is not likely to follow Viking to Mars, impact on the planet, and contaminate it. Thus, the two robots are set for an 11-month-long trip to Mars. But what exactly are these probes? While we have discussed what they can do, let's look more closely at how they do it.

[During nearly a year of interplanetary flight, the all important landing vehicle is in a dormant state, resting safely in a bioshield shell attached to the orbiter. This casing has prevented contamination during launch, and will stay with the lander until it reaches Mars. Meanwhile, the orbiter is doing all the work. It checks up on the landers health every 15 days and relays this data to earth. Occasional flashes from the orbiter's engines see to it that the proper course is maintained. Finally, the orbiter is responsible for an hour-long burn of its engines which slows down the two space laboratories so that they can be captured by the gravity of Mars. When this has been ac-

3   A parking orbit is that path around earth, the moon, Mars, or another heavenly body, from which a space craft launches itself or a ward vehicle on either a landing or interplanetary journey. The parking orbit is usually used as a time for checking that all systems are operative.

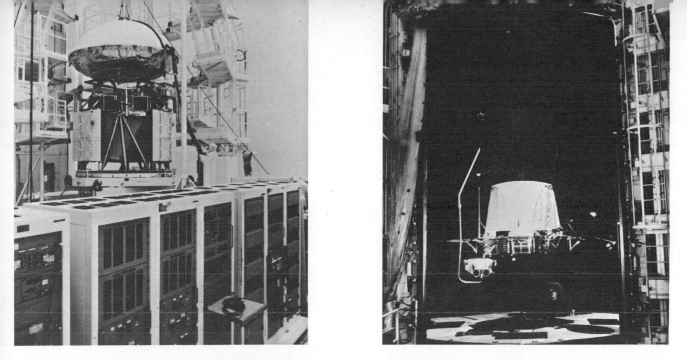

Left: The complete Viking model at the Jet Propulsion Laboratory. This model is used to qualify the orbiter/lander unit for whatever environmental conditions it may encounter. *Right:* The space simulation tests of the orbiter unit.

complished, the orbiter begins the search for a suitable landing site.

[The design of this octagonal nursemaid craft is based roughly on that of the Mariner, although it is considerably more sophisticated. With more duties to perform, the orbiter boasts two 4,096-word computers as opposed to the simple, specialized brain of its predecessor. It also has three propellant tanks which are three times the size of the ones which power Mariner and enable the four tons of Viking to maneuver through space. Attachments unique to the orbiter include venetian-blindlike shutters which automatically open and close to allow the 16 equipment bays to cool[4]; solar-energy reflectors to regulate the temperature of the propulsion unit; and insulated blankets to protect the propellant canisters. Both crafts boast four winglike solar panels, although the Viking appendages are nearly twice the size of those on Mariner. As a result, they are hinged in the center, which allows them to remain folded until the orbiter is released from the fairing of the launch vehicle. When fully extended, they offer a

4   The orbiter bays are as follows: radiofrequency and modulation/demodulation subsystems; central computer and sequencer; attitude control high-pressure module; data storage subsystem; attitude control and articulation control subsystems; flight data subsystem; visual imaging subsystem; Mars atmospheric water detector and electronics; batteries; power; attitude control high-pressure module; power; batteries; data storage subsystem; relay radio and relay telemetry subsystems and pyrotechnics; radiofrequency subsystem.

139

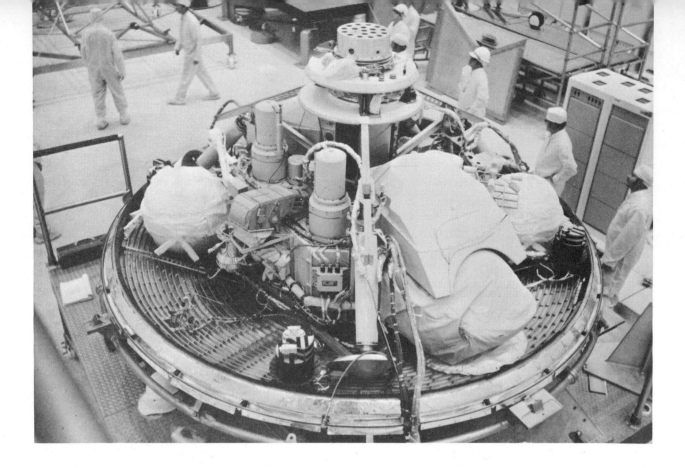

Above: At Martin Marietta in Denver, the Viking lander model undergoes drop tests to determine its durability. *Below:* The Viking lander capsule arriving at the Scapecraft Assembly and Encapsulation Building for the installation of flight subsystems prior to its baking and sterilization.

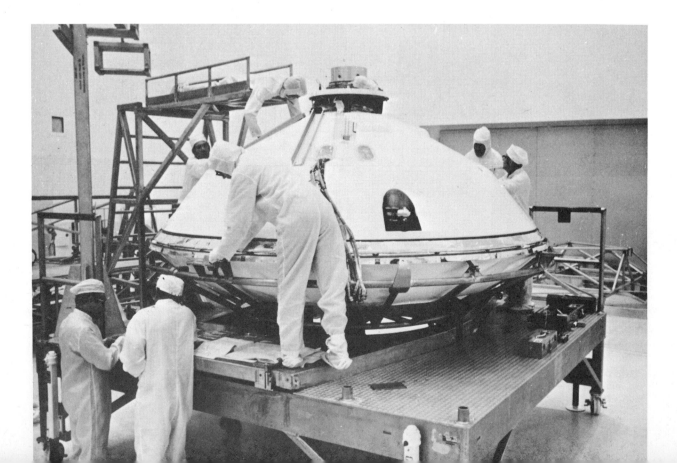

surface of 23,250 square inches of solar cells to the sun. As far away as Mars, they are able to supply the orbiter with 620 watts of electricity, supplemented by the power of two 30-ampere-hour nickel-cadmium storage batteries when the craft turns from Sol.

[Communications wise, Viking depends upon a parabolic high-gain antenna for effective receiving and broadcasting. This dish, 57.9 inches in diameter, is motor driven for precision beaming of all Viking data to earth. There is also a small, tubelike, low-gain antenna on the other side of the Viking. It cannot direct a signal earthward with quite the same pinpoint accuracy, but is useful when the high-gain unit is not facing in our direction.[5] A third antenna, located on the underside of one of the orbiter's solar panels, is used for lander-to-orbiter communication once the two ships have separated. In terms of power, the Viking's space-bound half transmits signals at 20 watts. Considering that most of our commerical radio stations operate at levels of over 100 times this amount, and with far less distance to cover, it is only the sensitivity of a 210-foot diameter Deep Space Network antenna that allows this transmission to be perceived.[6] And, on occasion, the orbiter literally floods these receivers with signals. During the photo relay sessions, for example, the orbiter's high-gain antenna sends home as many as 4,000 bits of information per second. Compare this with the eight and one-third bits from both Mariner and the Viking's low-gain antenna!

[Like the journey of Mariner, the flight of Viking is guided by the star Canopus. Using this stellar beacon for general orientation, the orbiter keeps both itself and the lander on course, making minor attitude adjustments with the one jet located on each solar panel. Drawing on nitrogen gas stored in a pair of 12-inch bottles, these rockets gently push the Viking into a proper trajectory whenever it

5 It should be pointed out that for the convenience of earthbound scientists, orbiter one can be "walked" around Mars so that it can contact and relay information from lander two, and that orbiter two can likewise communicate with lander one. Perhaps, if Mariner IX had not run out of fuel, it would also be a part of this amazing Mars network!
6 The Deep Space Network (DSN) is an international tracking and data receiving facility in which six 85-foot parabolic antennae and three 210-foot dishes relay outer space information around the world. In the case of Viking, the 210-foot antennae are most important: only they are sensitive enough to pick up the Mars data at the rapid rate with which it is being broadcast. Located in Goldstone, California; Tidbinbilla, Australia; and Madrid, Spain, these three paraboloids are so positioned that one of them is in constant contact with the Red Planet. Information from these dishes is sent to the JPL via cables, satellites, and microwave relays.

Left: The Viking lander being de-encapsulated after arriving at the Spacecraft Assembly and Encapsulation Building at the Kennedy Space Center in Florida. *Right:* Martin Marietta technicians work on the interior of the aeroshell. This 11-foot diameter structure protects the Viking lander as it enters the Martian atmosphere.

shows even a fractional deviation. And the result of these many terrestrial and mechanical operations? After crossing 440 million miles of space, the Viking package inserts itself into Martian orbit, circling the planet in an ellipse that brings it as close to Mars as 940 miles and as far away as 20,400 miles in a revolution which lasts 24.6 hours. Once in orbit, the process of surveying pre-selected landing sites begins. These were chosen from Mariner photographs, and the criterion used in finally approving a particular site is based on whether or not the location appears to be a safe, smooth place on which to set the lander. When this has been determined, the order goes out for the lander to wake up and spend approximately 30 hours becoming electrically independent of the orbiter. During these maneuvers, the fiberglass bioshield is discarded, leaving the lander encased in a protective, convex aeroshell. When these various chores have been completed, two small charges are detonated that release a pair of compressed springs located between the orbiter and the lander: fully expanded, the coils push the crafts apart. Some 10 minutes after complete disengagement, four hydrazine rockets on the lander's aluminum-alloy aeroshell are ignited and, as the orbiter continues along its path, the lander commences a deorbit procedure. With a cork-honeycomb coated surface facing Mars, to supply a small measure of lift and to protect the lander from the 2,700-degree F. heat

Left: The launch vehicle for Viking II. *Right:* NASA's Don Davis shows the parachute of the Viking lander package about to unfurl.

of entry into the atmosphere, the Mars lab begins its descent. During this final leg of the journey, the spacecraft is controlled by one 18,000-word Guidance Control and Sequencing Computer, although there is a second and identical computer onboard to serve as a backup system. The lander's instructions have all been recorded on the plated-wire memories of these twin mechanical brains, but they can be changed by a command from earth.

[Between two and five hours have passed and, at a height of 800 thousand feet, the craft reaches the outer fringes of the planet's air blanket. From this point forward, Viking is on a preprogramed, fully automatic schedule; instructions from earth would take 19 minutes to reach the craft, too long to be of any use as Viking lowers itself to the planet. Traveling at 1,230 feet per second, the lander's three, 18-nozzle terminal phase hydrazine engines fire and, a few minutes later, the Viking lander makes contact with the Martian surface at a comfortable 5.5 miles per hour.[7] Its three legs compress into the craft upon

7   The reason that the Viking thrusters have 18 nozzles per engine is to prevent a harsh scalding of the soil. Beyond the actual burning effected by these boosters, the propellants fired by three solid engines might have poisoned the soil and killed any microscopic Matians. Originally, the thought was to shut the Viking rockets 10 feet in the air and let the craft simply fall the rest of the way. Then it was discovered that using ultra-pure hydrazine fuel spread by numerous nozzles into a soft, fanning exhaust, would barely disrupt the surface of the planet.

impact, feeding against aluminum padding, which buffers the blow to the automated laboratory. Viking is now ready to begin its exploration of the Martian surface. However, the lander has not been taking it easy during its dramatic descent: as the first earthship to penetrate the atmosphere of the Red Planet, it had been performing experiments even as it entered the tenuous Martian exosphere.

[　　The six to 13 minutes that the lander spends in Mars' atmosphere are used to sample the air and make numerous scientific observations. The electrical processes of the upper atmosphere are the realm of the Retarding Potential Analyzer. As this device sweeps through the air, it is penetrated by ions and electrons. An electrometer measures the current of these particles. This tells how effectively ions from the sun (solar wind) are or are not deflected by Mars, and thus indicates the strength of the planet's magnetic field.[8] However, since most of the particles in Mars' atmosphere have no electrical charge, it is up to the Mass Spectrometer to determine what these gases are. And it does this by bombarding them with electrons and transforming the neutral atoms and molecules into ions. Electrostatic and magnetic fields in the instrument are thus able to pigeonhole the ions according to their atomic mass, and are then read by the Mass Spectrometer. Other atmospheric tools onboard Viking are a Recovery Temperature Instrument which, as its name implies, measures the temperature; a Stagnation Pressure Sensor Cell, a pressure-gauging device; and Accelerometers, which tell us about the density and aerodynamic traits of Mars' atmosphere. All of this information is stored on tape for eventual broadcast to the orbiter and subsequent relay to earth.

[　　Because the data gleaned by Viking is so important to scientists, the craft's first job after landing is to establish communication through the orbiter with the terrestrial DSN antennae. This system is not a perpetual one: the orbiter can shuttle information from the lander to earth only when it can be seen by the grounded craft at least 25 degrees above the horizon, and no farther than 3,000 miles away. This link is accessible for a period or *window* of 10 to 40 minutes each day, although direct contact with the lander can be had when a DSN 210 foot dish is perfectly aligned with the lander's own high-gain antenna, a 30-inch diameter unit located atop the vehicle. In either case, due to the Viking's limited electrical power, we can talk with the craft

8 Sunlight, as opposed to solar wind, dissociates carbon dioxide to create the ionosphere.

for no more than two hours each day. During this period, the lander-orbiter-earth link can supply us with 16,000 bits of information per second; the earth-lander union can handle only 500 bits per second.
[In order to inform the earth of its safe descent and various scientific findings, the heat from the decay of the element plutonium 238 is converted into 70 watts of electricity by the lander's Radioisotope Thermoelectric Generator and is used for transmitting as well as for all other power. Resourcefully, the heat that is not immediately used by Viking is transferred to a storage compartment for later use, rather than being thriftlessly thrown into the atmosphere. Given the 50 percent ratio of the sunlight received by Mars as opposed to earth, and the fact that the surface temperature on the Red Planet drops to as low as −184 degrees F., this salvaged heat is indeed valuable. Within a few minutes of landing, then, Viking is busy reviewing its various components and familiarizing itself with its new home. All of the scientific tools and limbs have been extended from the craft's hexagonal body—which is 58.8 inches wide and 18 inches deep—and, when the initial report is concluded, the lander goes obediently about its tasks.[9] Not surprisingly, the first of these chores is picture-taking. At long last scientists will know what has been enthusiastically debated over the centuries: what does the surface of Mars look like? Ten minutes after landing, one of the probe's two stereoscopic cameras begins to obtain visual or so-called *imaging* data.[10] The cameras are so positioned that they can photograph the entire 360-degree sweep of the Viking lander site; horizontally, they can pan from the vehicle's footpad to approximately 40 degrees above the horizon. This wide range is made possible by a set of mirrors that move with the cameras to complement their view. However, the second camera is not called into operation until the local Martian weather has been determined: a sandstorm, for example, could easily obliterate the carefully ground Viking lenses. These pictures, then are

9 A great deal of thought was given to the actual positioning of the Viking experiments so that one would not interfere with another. For example, the vents for the life-search experiments could not be placed near the inlets for other tests, lest an expelled gas be drawn in and mistaken for typical Martian atmosphere. Too, experiments had to be scheduled so that, to cite one instance, the Viking engines would be off while the seismic experiments were being conducted. Even the light vibration of the craft might be mistaken for a Martian tremor!
10 The cameras were so positioned as to approximate the ratio of the distance between human eyes. Since the slightly different perspective seen by each eye is what gives us a three-dimensional view of the world, the location of the Viking eyes gives us a stereoscopic view of that world.

146

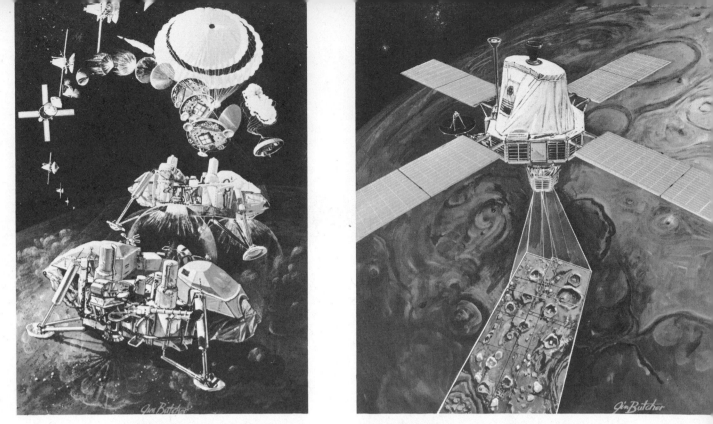

Left: Illustrator Jim Butcher's interpretation of the Viking landing sequence. *Right:* After the lander has separated, the Viking orbiter uses its two television cameras to shoot overlapping pictures of Mars, which it transmits to earth.

intended to provide immediate food for the hungry geologists and biologists to digest, while giving mission directors an opportunity to study the terrain from which the lander's scoop will dig soil samples for automated chemical analysis.[11]

[The taking of these surface photographs is not radically different from the Mariner dot or *pixel* method. Light enters the camera lens and falls on photosensors, each of which generates a signal relative to the intensity of the light that is striking it. This data is then broadcast to earth, decoded, and recorded on film. The only difference between the Mariner and Viking transmissions is that each pixel sent from the lander is more intricately described in a seven-bit rather than a six-bit word. However, of considerable scientific value is the fact that color shots can also be taken by Viking, using three filters through which the same scene is recorded thrice in succession. Each filter reduces the landscape to a code corresponding to red, green (yellow), or violet (blue), the tones from which all others are derived. These separate shots are then reconstructed in color on earth, and combined to render the vista in *full* color. Since the known color of

11 The nine lander scientific groups are as follows: imaging, seismology, biology, entry, molecular analysis, magnetic properties, meteorology, inorganic chemical, and physical properties. The orbiter teams are imaging, water vapor mapping, and thermal mapping.

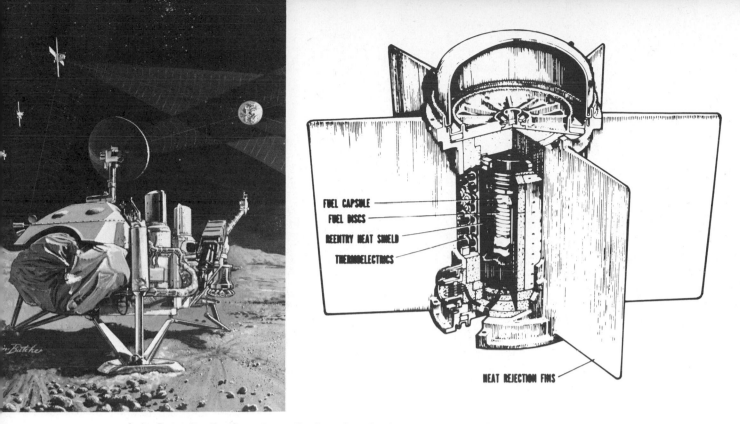

FUEL CAPSULE
FUEL DISCS
REENTRY HEAT SHIELD
THERMOELECTRICS

HEAT REJECTION FINS

Left: Artist Jim Butcher pictorially describes the lander-to-orbiter-to-earth radio link. *Right:* A NASA rendering of one of Viking lander's two 35-watt radioisotope thermoelectric generators.

certain portions of the spacecraft will be visible in most any pictures, imaging personnel are able to jockey any discrepancies in tone until the true hue is achieved.

[Beyond the photographic investigations of Mars, the Viking lander has the task of searching for life, analyzing the weather with temperature, pressure, and wind sensors, measuring Mars-quakes with a seismometer, seeking out magnetism with two magnet arrays and a magnifying mirror, and dissecting the soil with a Gas Chromatograph and an X-ray Spectrometer, and the atmosphere through a Mass Spectrometer. Let's look at these experiments individually.

[We've already met the Mass Spectrometer during the Viking entry stage, and the procedure followed by its land-locked operation is no different. The Gas Chromatograph and X-ray Spectrometer, however, are strictly tools for surface exploration. Unlike the biological experiments, which search for living organisms, these comparatively incomplex units look at the chemical nature of the Martian soil. Since chemical evolution must precede biological growth, the elemental state of the planet's crust has much to offer scientists in their search for Martian history and life forms. The first step in this study is to grab a sample of soil using a derricklike scoop, a device we'll look at in detail when we study Viking's direct quest for life. The handful of

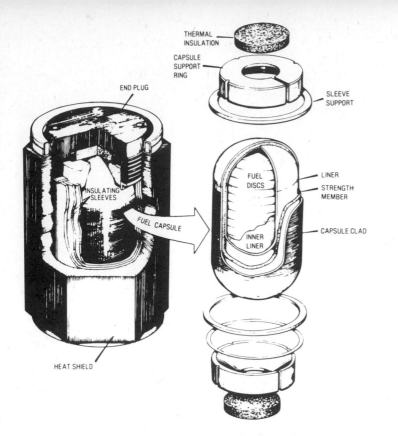

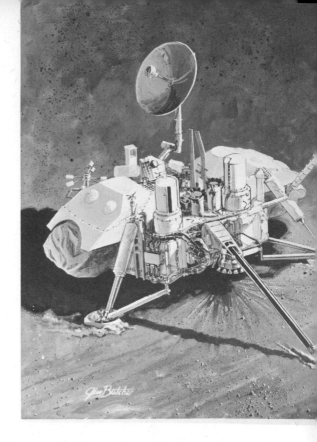

Left: A closeup of the fuel capsule which powers Viking's radioisotope thermoelectric generator. *Right:* Jim Butcher's pre-landing sketch of Viking on Mars.

Martian terrain is then placed in a small oven that is part of the Gas Chromatograph. This heating process vaporizes inherent organic (carbon-based) compounds which, as we saw in chapter two, are the precursors of life. These gases then pass through a chromatographic block and are separated, since different organic gases travel at varying speeds through the column. Thus, as the gases creep through the other side of the Chromatograph, they are an easy mark for identification by a waiting Mass Spectrometer. Contrarily, the X-ray Spectrometer studies a sample of Martian soil to identify inorganic compounds: in other words, the Martian minerals. We know a great deal about the sands and rocks that make up the earth and moon; information of this sort about Mars will fill in the questions geologists have regarding the planet's origin and composition. Quite simply, the scoopful of Mars is dropped through a funnel into a chamber where it is subjected to the X-rays of radioactive iron-55 and cadmium-109. This causes the soil sample to become flourescent, each individual element radiating X-rays that are characteristic of that particular element. This entire process takes approximately five hours. Another exploration of the Martian surface is conducted by magnets attached to the soil scoop where iron-bearing or magnetic particles will be drawn to them; here they can be photographed in color through the

150

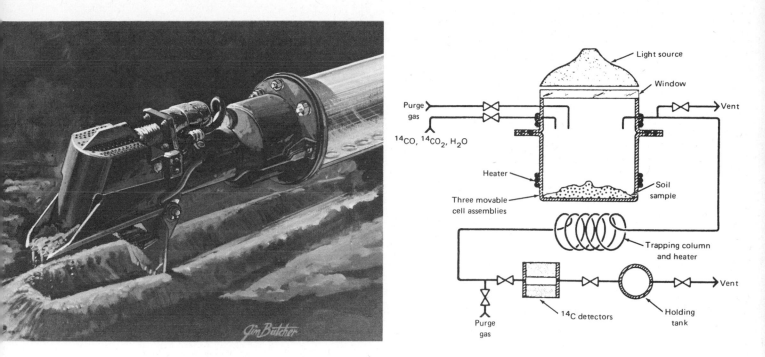

Left: In this Jim Butcher sketch, Viking's scoop samples the Martian terrain. Drawings such as these were released to the media to explain the workings of the interplanetary laboratory. *Right:* The Pyrolytic Release Experiment.

four-inch magnifying glass. These pictures will help us to recognize the various minerals which Viking uncovers, and thus tell us something of the planet's past in terms of oxidation and material differentiation (such as occurred on earth). Yet, while these tests are interesting, more *dramatic* is the seismometer which detects the Marsquakes. A tremor of any significance will travel up the lander's three legs to the instrument, which will then study the information and beam it earthward.

[Once again, as we will see when we examine the search for Martian biology, the tests we have just examined are relatively uninvolved. Yet, even less elaborate are the workings of humankind's first Martian weather station. Located on a hinged boom which juts from the lander body, all of the meteorological gadgets are of the sort we use on earth. On top of the appendage is a small, fingerlike extension known as a Hot-film Anemometer. Like a terrestrial anemometer, the 2/5 inch long tool measures the wind. Inside the protective shell of aluminum oxide is a thin platinum film that is heated by electricity. The faster the speed of the wind, the more the platinum layer is cooled. The other environmental units are a Stretched-diaphragm Sensor to chart pressure, and the temperature-detecting Thin-wire Thermocouple. The pressure gauge is nothing

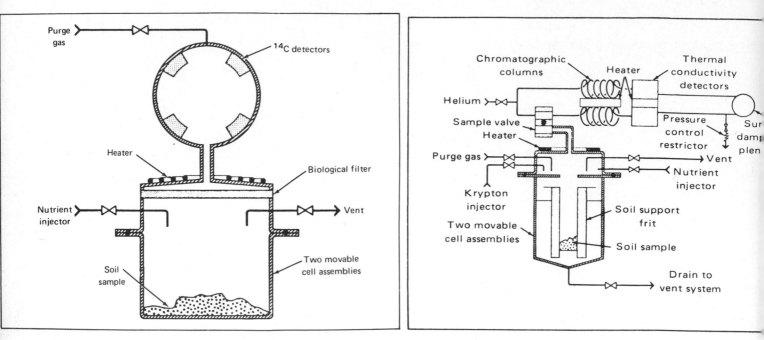

Left: The Labeled Release Experiment. *Right:* The Gas Exchange Experiment.

more than a capacitor, a device which gathers electricity. Since a change in pressure will alter its effective level of capacitance, the unit can interpret pressure as it is expressed in current. The Martian thermometer also works on the principle of conductivity, in this case a strip made of two dissimilar metals bending as the temperature increases or decreases. Because these two metals react differently to heat and cold, the degree to which they "give" respective to one another tells us what change there has been in the temperature. Both the capacitor and the thermocouple are exposed to the air, although the fragile thermocouple is backed on one side by a supportive wind shield.

[With the exception of the photographs, and the feat of actually landing a robot on Mars, none of these Viking operations can arrest the imagination as absolutely as the search for life on the Red Planet. And why should they? We've had 80 years of conditioning! Ever since Wells and Lowell first singled out Mars as the likely abode of one form of creature or another, humankind has all but regarded Mars' fertility as a *fait accompli*. And media programing—the books, the movies, the scientific pronouncements—has kept this faith alive throughout the years. Thus, no one will be surprised to find life on Mars, although such a discovery may be accompanied by some measure of cultural

Left: Flying at 73,000 miles-per-hour relative to earth, Viking I photographed Mars from 7 million miles away. The small white bubble is an error in data transmission. *Right:* The launching of Viking II.

shock (see chapter seven). But what about the flip side of this celestial coin? What if there is no life on Mars? Scientists regard it as a no-lose proposition. In a general sense, the goal of science is to understand our universe through knowledge.[12] Whatever the results, we will certainly have gained an awesome amount of that commodity. More specifically, we will see that either the earth is uniquely vital, and try to find out why, or we'll learn that life has evolved elsewhere, thus giving us an invaluable comparative study for the understanding of interstellar biological activity. As capsulized by Dr. Philip Morrison, a physicist who works with NASA on projects involving extraterrestrial life, the identification of life on Mars would change the phenomenon from being a miracle to a statistic, a welcome perspective in any scientific endeavor. Of course, not finding life on Mars will not mean that life does not exist on Mars (we may land in the wrong spot) or elsewhere in the universe; again, the various forms and reality of extraterrestrial life will be discussed in chapter seven. Perhaps the only negative result of a fruitless life search will be to

12 The "understanding" aspect of science is generally referred to as pure science, and it is the most difficult kind of study to sell to the public. The more palatable programs are those which have widely affective goals in mind, be they positive, such as curing cancer, or negative, such as building nuclear weapons.

Above: Mars from 348,000 miles. Photograph taken by Viking I, with several of the planet's huge volcanoes visible in the picture. Olympus Mons is visible toward the top of the frame. At the bottom of the disk, the impact basin Argyre is plainly visible. *Below:* Mars from approximately 400,000 miles. The huge rifts known as Vallis Marineris or the "Grand Canyon" of Mars are visible near the terminator, just above the center of the picture. Argyre is just below the center of the photograph.

fuel the cause of those people who ask the tired question about why we bother with space. Hopefully, we will be wise enough to see beyond their short-sighted arguments.

[As one can imagine, looking for life on another world poses many tactical problems. For one thing, there is the scientist's aforementioned nightmare: that biology on Mars is geographically restricted, and that we set down in a dead area of the planet. Then there is the consideration that, unless we are dealing with a macrobe—a creature that can be seen with the unaided eye—the search for life can be unimaginably difficult, even if we've landed on an ideal location. Since Viking is not equipped with a microscope-camera unit—the cost would have been prohibitive, although the rewards would have been great—it must rely on other means of life detection. Scientists puzzled over the problem for 20 years before deciding upon three small laboratories which constitute Viking's biological hunt: the Pyrolytic Release Experiment, the Labeled Release Experiment, and the Gas Exchange Experiment. The one drawback to these projects is that they succeed or fail based on whether or not Martian biology has a metabolism (energy and growth processes) that is akin to the proverbial "life as we know it." Scientists are betting on the fact that a planet that was formed at the same time as the earth, is warmed by the same sun, and is so close to us in terms of distance and general composition, will host at least a close or recognizable relative to terrestrial life.

[The operation of these three experiments depends upon the depositing of soil samples within their respective chambers. And serving the scientific triumverate, as it had the search for organic and inorganic molecules, is a simple but functional scoop. Set between the two cameras, the soil sampler consists of a collector head which is driven into the terrain by an extendable boom. It can swing in a wide arc and reach as far from the craft as 10 feet. Snaring an interesting shovelful of terrain, it then retracts, moves to the appropriate funnel, and divests itself of the sample. In the Pyrolytic Release Compartment, the Martian environment has been recreated in every respect, including simulated sunlight from a xenon lamp. The only difference between it and the actual Martian atmosphere that is admitted through a vent is that the indigenous carbon dioxide is replaced with carbon monoxide and carbon dioxide spiked with radioactive carbon-14. When the .02 in.^{3} soil sample has been in this setting for at least

155

The working model of the Viking lander, after its various tests to substantiate the validity of its design, is used by the Jet Propulsion Laboratory to duplicate the maneuvers being executed on Mars. Through this working model, technicians were able to solve the various problems that arose.

five days, the chamber is flushed of any residual carbon-14 and is heated to 1,160 degrees F. This causes any organic compounds which may be present in the sample to "crack" and release their vapors. These are then searched by special monitors for traces of radioactivity. If the Martian microbes act anything like earth biota, they will have assimilated the carbon-14 by "breathing" the air in their normal metabolic activities, and the tracer will show up after the pyrolization process. This transfer will indicate the presence of biological chemistry. Viking is equipped to repeat this experiment up to four times. The modus operandi of the Labeled Release Experiment is quite similar to that of the Pyrolytic Release. A total of .03 in.^{3} of soil is fed to the chamber which is equilibrated with the Martian atmosphere and sealed. This time, however, instead of tagging the air, carbon-14 is injected into a nutrient mixture of organic and amino acids, with which the soil sample is nourished for 11 days. If, after this period, the carbon-14 is found in the atmosphere of the canister, then it is clear that a breakdown of the acidic foods has occurred, evidence of inherent biology. This experiment can also be performed numerous times. Finally, there is the Gas Exchange Experiment, which measures the changes which occur in an atmosphere composed of the metabolically crucial gases carbon dioxide, nitrogen, oxygen, methane, and hydrogen. The unlit cell is given .06 in.^{3} soil and is closed, the sample washed with a rich nutrient, and, after several days, a gas chromatograph reads the atmosphere. If the balance of gases has changed, the inference is that living organisms are present in the soil. Like the other experiments, the Gas Exchange can be worked many times over.[13]

[These, then, are the many scientific experiments of the Viking lander. And while all of this is going on, of course, the orbiter is still performing its observations from up above. In fact, in the case of recent space history, the Viking I orbiter was busy scouting sites for a Viking II landing while that second craft was en route to Mars. Obviously, the Viking is simply a mind-boggling advance in space science. Thus, with a general understanding of the equipment and the planet itself, let's turn to the missions of Viking I and II and see what

13 As of 1970, the three Mars lander life-search tests were the Gulliver experiment, in which carbon dioxide was laced with carbon-14 and fed to a soil sample through a liquid nutrient; the Multivator, wherein surface samples would be tested for reactive materials; and the Wolf Trap, an experiment in which the atmosphere was inhaled and tested for signs of airborne biota.

Above: Perhaps the most extraordinary photograph ever taken, this was the first shot taken by Viking I after its landing on Mars. The center of the picture is approximately 5 feet from the camera. Dust kicked up by the lander's descent engines has settled in the dish of the spacecraft's footpad. *Below:* The first color photograph of the surface of Mars. The reddish tint of the sky is due to reflection caused by the red-colored dust hanging in the lower atmosphere. A second photograph, of the lander's color chart, with its known hues, was used to balance the values and intensity of the colors in this shot.

they have told us about the most alluring planet in our solar system.

[The afternoon was clear and the sky a deep blue when, at 5:22 P.M., Eastern Daylight Time, on August 20, 1975, the Titan/Centaur rocket housing Viking I was launched into space. Twenty days later, on September 9, 1975, at 2:39 P.M., EDT, Viking II went roaring skyward atop twin torches of bright orange flame and rolling white smoke. It was planned for the first Martian lander to set down on July 4, 1976, in honor of the American Bicentennial, and for the second Viking to make its descent within a month after that.

[Things didn't quite work out that way.

[Viking I arrived at Mars right on schedule, on June 19, 1976. The orbiter fired its 300-pound thrust engine for 38 minutes, and went sailing into a wide, elliptical orbit around the planet. Almost immediately, the orbiter's cameras began sending back a number of surprising photographs, such as those showing previously unseen patches of frost or fog on the planet's surface, enhancing the possibility of finding water and, therefore, life. However, Viking also revealed a problem with the prime landing site (A1) in Chryse Planitia. The Chryse basin had been selected because it was revealed by Mariner 9 photographs to be the juncture of several channels which may once have been rivers. Scientists did not know whether the water had evaporated, retreated to the polar caps, or was frozen beneath the Martian crust. But they did feel that if there is life on Mars, it must exist in the once-moist-sediments of these long-dead troughs. Unfortunately, the orbiter pictures showed the Chryse to be much more rugged than had originally been thought. Since landing on a fair-sized boulder or on a crater, hill, or channel with an incline of over 19 degrees could damage or upend the craft, a safer landing site was sought. The task is not an easy one. Viking is on an aforementioned automatic sequence once it separates from the orbiter, and it must be given a large, oval area some 60 by 160 miles in which to set down. Specific sites cannot be targeted, nor can the orbiter see anything smaller than 90 yards across. Thus, after the order is given for the lander to descend, a safe setdown is the work of chance!

[At this point in the mission, project directors had two alternatives. They could compromise with the terrain and seek out the safest area bordering the primary locale, or they could change the craft's orbit entirely and shoot for an alternate site (A2) in Tritonis Lacus, which was several thousand miles from Chryse. Since the A2 landing

159

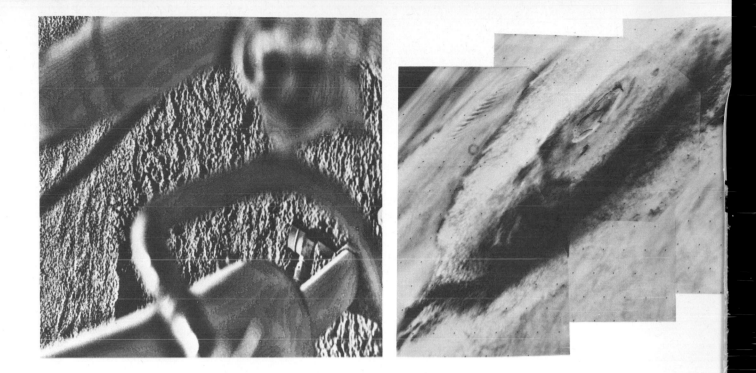

Left: One of first Viking's three footpads, which should be visible in this picture of the lander's leg, is buried beneath a loose cover of Martian soil. *Right:* From 5,000 miles up, the Viking I orbiter took this extraordinary picture of Olympus Mons, the awesome volcanic crater. Reaching into the stratosphere of Mars, the landmark is surrounded by cloud. This illustrates the atmosphere through which the lander plunged.

would have required complicated orbital adjustments, scientists decided to stick with Chryse and found what appeared to be a smooth area 180 miles northwest of A1. July 4 came . . . and went. So did the second target date of July 17, when the new area was surveyed and found to be too rough. Then, south of this spot, a satisfactory site was found and, in the small hours of the morning on July 20—fittingly, the seventh anniversary of the day when humankind first walked on the moon—the Viking I lander was told to begin its trip to the Martian surface. Henceforth, all that JPL personnel could do was to sit, bite their fingernails . . . and wait.

[Three and one half hours later, at 5:12:07 A.M., Pacific Daylight Time, the incredible robot was safely on Mars, at longitude 47.5 degrees west, latitude 22.4 degrees north. Word of the landing that had occurred 212,000,000 miles away reached the earth 19 minutes later. "Touchdown!" cried an exuberant Richard Bender, one of the flight controllers. "We have touchdown!"

[To say that the people at the JPL, NASA, their many subsidiaries and contractors, and the DSN facilities were ebullient is an understatement. At the JPL the champagne was flowing freely; many of those who had worked for eight years designing, building, and finally flying the probe were in tears. Others were numb, quite frankly in

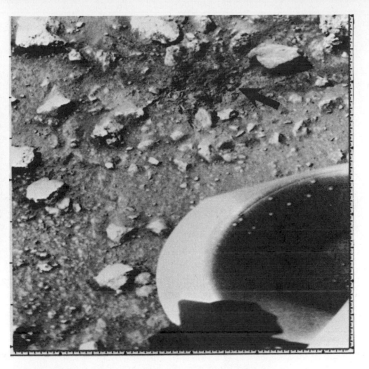

Left: A two-picture mosaic taken by the Viking II orbiter showing the volcanic plateau Alba Patera. This site was considered, early on, as a possible location in which to set down the second lander. Is Mars still a volcanically active planet, or is it dormant? Future probes will tell. *Right:* Another shot of the footpad seen in the historic first photograph illustrates a patch of dark material stirred up when the protective cover of the scoop was jettisoned, hit the ground, and bounced out of camera range. This photograph was taken two days after the landing of Viking I.

awe of what they had done. In a field such as space exploration, it's often possible to lose sight of the generic romanticism when one is so close to hard technology. But when the mind realizes that dream and imagination have become fact, the effects can be pleasantly jarring! Then, within 90 minutes of landing, came the photographs, the first one showing the lander's footpad and having been taken immediately upon Viking's arrival, and the second, a panorama of the horizon, shot six minutes after touchdown. They appeared on JPL television screens, one after the other, line by dramatic line, as the bits of information were decoded. "The details are just incredible," said Dr. Thomas Mutch, head of the Viking imaging team. "It just couldn't be better." "I think it's great!" Mission Director Thomas Young agreed. "It has exceeded all our expectations." President Ford phoned his congratulations: "I think it's amazing," he said, "that in the span of a single lifetime, the exploration of space has grown from the dreams of very, very few individuals to such a massive cooperative reality." It's a pity that Lowell and Goddard, the two greatest visionaries, could not have been there to share in the triumph.

[Even while the celebrations were ripe, however, curiosity was also rampant, and the various team members turned to Viking whenever communications allowed. By the time of the press confer-

The two photographs on the left: In preparation for the Viking Mars landing, the Jet Propulsion Laboratory prototype goes through the motions of its space-worthy counterpart. *Right:* A trench dug by the Viking I scoop in its search for life. The excavation is two inches deep, three inches wide, and six inches long. Area around the trench has ripple marks produced by the wind; the adhesion of the grains show that they behave somewhat like moist sand on earth.

ence at 10:30 P.M., PDT, July 20, scientists had had a chance to make a cursory study of the pictures and data which Viking had sent to earth. Dr. Young was the first to speak, and told newsmen that the lander was sitting on a spot that was slightly inclined by 3 degrees, and that its cameras could see in an arc of 322 degrees. The "gentle slope," as he called it, on which the craft was resting had an azimuth of 285 degrees. Young also explained that the first color picture of the Martian surface would be in and color-reconstructed by 8:55 A.M. the following morning. The next speaker was Dr. Alfred Nier, leader of the Entry Science Team, who explained the operation of the upper atmosphere experiments, although, at the time, he had no findings of which to speak. Al Seiff gave a brief talk about the tests in the lower atmosphere, but he, too, conceded that "it is too soon to make a detailed report." For, while Viking had radioed back a torrent of information, "we have had, at this point in time, only the opportunity to do manual calculations, sitting and punching buttons on the calculator and sorting numbers." During the coming day, a computer would be brought to bear and, Seiff promised, "within about 24 hours, we will have a fairly complete definition." However, Seiff did have some general findings to report, which he himself found "interesting." The first was that the surface pressure at the landing site was

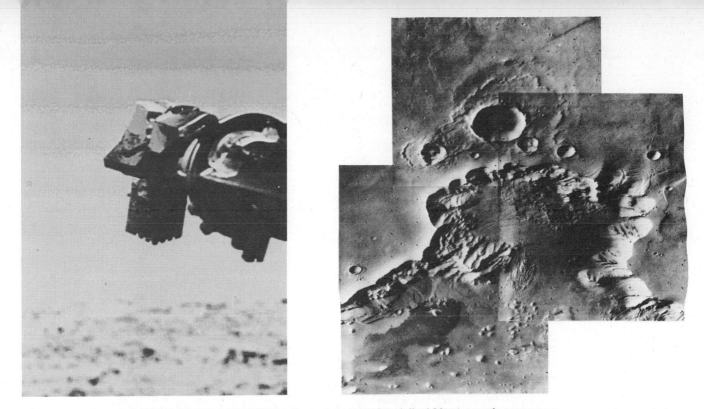

Left: The collector head of Viking I's surface sampler, full of Martian soil, prepares to carry its load to the gas chromatograph mass spectrometer. *Right:* A field of sand dunes *(lower left),* some 30 miles long, and a massive avalanche can be seen in this Viking orbiter shot of the north wall of the Gangis Chasma.

discovered to be roughly 7.3 millibars, a millibar being the standard measure of atmospheric pressure. For comparison, the pressure at the earth's surface, the mean figure of which is water level, is 1,014 millibars. "Now," Seiff deduced, "since the accepted value of the surface pressure (on Mars) in the last two or three years . . . has become 6.1 millibars, we can . . . say what the elevation is of the landing site of the spacecraft, and it turns out to be 1.74 miles below the mean figure of the planet." Subsequent reinterpretation of the data refined the reading to be 7.7 millibars at an aproximate depth of 1.8 miles.[14] "With respect to the temperature," Seiff went on, "we found that the temperature at landing was a number in the neighborhood of −60 degrees F. While that seems rather cool by earth's standards, it is somewhat warmer than many of the previously existing models of the atmosphere of Mars." Additional readings put the temperature range at a low of −122 degrees F. just after Martian sunrise, and a high of −22 degrees F. later in the day. The entry temperatures covered a somewhat broader spectrum, reading 27 degrees F. at the outermost

14 Prior to Viking and Mariner, scientists had estimated the pressure on Mars to be between 25 and 80 millibars. This guess was based on the way that light was scattered about the atmosphere, although it was discarded when the dispersion was found to be the result of great clouds of dust.

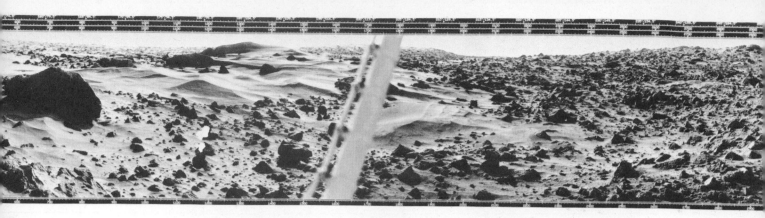

One of the most spectacular photographs to come from Mars: 100 degrees of the Martian terrain, taken during the sharp, early-morning light. Viking I's meteorology boom, which supports the vessel's miniature weather station, is seen in the center.

fringes of the Martian air blanket, and dropping to −216 degrees F. at 81 miles, where there was found to be a heavy ion concentration.[15] Seiff closed his briefing by noting that "with hand computers, we couldn't tackle the problem of the wind, yet," but wind speeds were later determined to be 15 miles per hour, coming from the east in the afternoon of the first day, and issuing from the southwest by midnight. The remaining speakers all echoed what Seiff had said, agreeing that there would be more substantive information to reveal in subsequent gatherings.

[For the next few days, as scientists struggled with the brief communication windows, a great influx of data, and their various computers, the press conferences were generally uneventful. Typical was the meeting held on July 21, again at 10:30 P.M. It began with a general post-landing status report from Deputy Mission Director Lou Kingsland. "I am happy to report," he began, "that the Viking lander has now spent two successful nights on the surface of Mars and is beginning its second full day of operation. Very early this morning, very shortly after midnight, we were able to communicate directly with the lander on the surface of Mars from out earth stations, and I'm happy to report that we were able to successfully communicate both with commands going up to the lander and to successfully receive telemetry coming down from the lander. All indications are that the lander weathered the first night on the surface in very good working order and that it should continue to do so indefinitely. Later this morning, we received a large quantity of information relayed from the lander via the orbiter to earth from about 8 to 10 A.M. this morning, and

15   There are two types of ions: a cation, or positively charged ion, in which an electron has been lost; and an anion, a negatively charged ion in which an electron has been gained. An ion can also refer to a charged subatomic particle. Clearly, the effect that an ion will have on its surroundings depends on exactly what type of ion it is.

164

Visible in this broader panorama of the same terrain covered in the 100-degree photograph are the two United States flags located on the dual Radioisotope Thermoelectric Generator wind screens, as well as the meteorology boom and, between the flags, the support struts of the high gain antenna. Notice clouds just above the horizon.

we received about five pictures and a significant quantity of other information and data. The relay-link performance that we're getting from the lander to the orbiter has exceeded our expectations to the point that we now feel that it will be possible to greatly expand the quantity of science data eventually collected from the lander and played back to the earth over that relay link. We feel very happy about that result. The orbiter continues to operate normally; the second spacecraft [Viking II] continues operating normally on its way to Mars." There was, in fact, only one problem with the Martian lander as Kingsland explained in this lander review: "The telemetry indications are that the lander seismometer is not working properly. Preliminary analysis indicates that the sensing elements in the seismometer—there are three of those—did not uncage after landing. The elements are caged at the time of landing to prevent damage due to landing shock. The commands that are designed to uncage the seismometer are stored in the lander computer and are issued automatically by the lander computer to the seismometer—or were *supposed* to have been—about four hours after touchdown. An analysis of the relay data that we collected earlier today indicated that apparently the seismometer did not uncage. An engineering team composed of members of the Viking Flight Team and additional experts from around the country is now being mobilized to try to understand the problem and to determine what the cause and the possible cure for this anomaly might be. Reasons for this failure include the possibility that the computer onboard the lander may not have issued the correct command to the seismometer. This is not considered very likely to have happened. It's also possible that something in the electrical circuit onboard the lander has broken. It's possible that the arming plug, which is involved in the circuitry for firing the uncaging device in the seismometer, is either broken or is

Left: Visible in this photograph are the United States flag, which is printed on the Radioisotope Thermoelectric Generator wind screen, along with the bicentennial logo and the Viking symbol. The Viking logo was created by Peter Purol, a Baltimore high school student who had entered the design in a NASA contest. To the right is the Reference Test Chart used in balancing the colors of the photographs. *Right:* The small "x" in this Viking II orbiter photograph marks the eastern end of the Utopia Planitia, the landing site of Viking II.

not working. It's also possible that the mechanical aspects of the caging mechanism in the seismometer did not work." Other than that, Kingsland was pleased to report that everything was in order, and over succeeding weeks, prior to the activating of the life-search experiments, the following facts were made public at JPL press conferences.

[The soil of Mars was discovered to be rich in iron (12–16 percent), silicon (13–15 percent), calcium (3–8 percent), aluminum (2–7 percent), and titanium (.5–2 percent). This suggests that the reddish color of the rocks and soil, as revealed by the color photographs taken by Viking, is due to a thick coating of the iron ore limonite, which is not much different from the terrestrial situation known as desert varnish. Above the rock-blanketed terrain is a pink sky, the result of sunlight being diffused by airborne dust particles. Earth's blue sky, of course, is caused by the way our atmosphere scatters sunlight toward the blue end of the spectrum. As for the air on Mars, while we discovered only a sparse sweeping of solar wind, the planet has an abundance of carbon dioxide, nitrogen, and argon, mostly in the upper atmosphere, with detectable amounts of ozone, between 0–5 miles of the surface. And although the quantity of the gas is not sufficient to keep the planet from being baked with ultraviolet radiation, the bombardment

166

On the left-hand side of this photograph is a dune created by the digging of the soil sampler. The scoop can be seen in its temporary "park" position. Regarding the geology of this vista, the rocks are of various types. Some are coarse-grained and knobby, such as the 10-inch-long rock just to the right of the trenches, while a more common type is the lighter-colored, irregularly shaped, coarse-grained sort such as the one seen just above the scoop. The lighter-colored patches in the middle and far field are outcroppings of the bedrock which underlie the landing site. Large blocks, 3 to 6 feet across, can be seen on the horizon about 330 feet from the spacecraft. It is possible that these rocks lie on the rim of a crater that forms the horizon. The cracked surface near the lander was caused by its rockets. Mounted on the end of the boom at the lower left is the magnet-cleaning brush, while the box in the lower right corner housed the meteorology instrument prior to its deployment a few minutes after landing.

occurs in the 2,000 angstrom range. Since the energy levels are inversely proportionate to the angstrom levels, as one scientist put it, "I would not consider the radiation environment as in any sense negative on the question of biology." Naturally, the discovery of this gas is, in itself, a positive sign that there is or was, at one time, some form of life on the Red Planet. However, there are other criteria which more closely relate to the life processes than the presence of ozone. There is, for example, the question of the makeup of the Martian atmosphere close to the surface of the planet. And with Viking I, we were beginning to see more precisely than ever what elements composed the air, as well as their ratios—although interpreting these distributions was quite another matter. According to early Viking tests, the atmosphere of Mars contains 1.6 percent argon (50 percent more than the earth) in the form of its isotopes argon-36 and argon-40. The significance of this finding was considerable. There had long been speculation among scientists that Mars is experiencing an ice age. The water would thus be locked beneath the planet's surface or at its poles, while the atmosphere would be drawn to the poles and frozen, which would account for its frail constitution. However, argon, being an inert gas, would not be subject to this freezing, or other such metamorphoses like combining with the soil the way oxygen does.

Left: While the lander was busy on Mars' surface, the orbiter shot such pictures as this one, of an area west of Argyre in the southern highlands. The channels may have been carved by streams which once coursed over Mars. *Right:* The Viking I orbiter snapped the four photographs which form this montage. Visible from a height of 4,100 miles is the huge Arsia Mons, a volcano over 62 miles in diameter and 17 miles high.

Thus, the amount of argon in the atmosphere has always been constant; from these measurements, we can therefore construct the nature of the Martian atmosphere as it existed hundreds of millions of years ago. It's analogous to visualizing the size and weight of a man by examining his footprints. And the concensus regarding the Martian air blanket is that it was once much thicker and thus conceivably more supportive of both water and life. Abetting this favorable finding were the discovery of nitrogen, necessary for nitrogen fixation—the process by which plants and bacteria build amino acids, proteins, and other cellular components—present in 2.7 percent in contrast to earth's 78 percent[16]; .15 percent oxygen as opposed to the earth's 21 percent; and a whopping 95 percent carbon dioxide vs. the .03 percent on earth. The confirmation of this vast store of carbon dioxide on Mars was a double-edged sword in terms of the question of life: it means that carbon and oxygen, elements which are vital to the existence of living things, are present on Mars; however, it also indicates that the carbon dioxide-to-oxygen process of photosynthesis, the means through which plants build organic matter, is not taking

16 The importance of finding nitrogen was discussed by Dr. Nier: "Nitrogen was never detected by any of the remote measurements, either earth-based or made from satellites, and, of course, it is of extreme interest in connection with the whole question of biology. While there are some people who feel that you could have biological organisms without nitrogen, on the other hand, the fact of the matter is that nothing on earth that's living exists without the existence of nitrogen incorporated in the molecules."

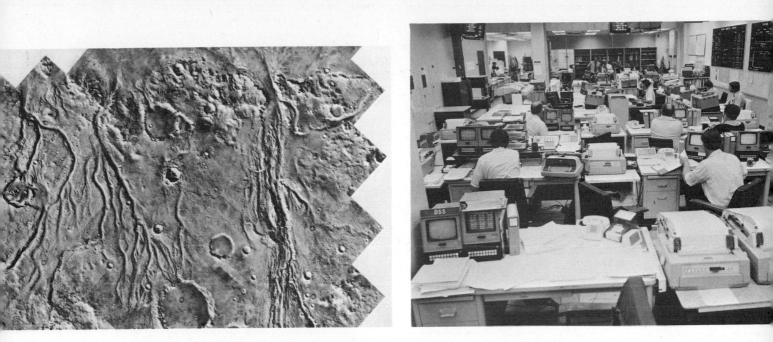

Left: These bold troughs indicate that this area to the west of the plains of Lunae Planum were once heavily flooded. In several cases, the channels cut through craters and, in other instances, the craters clearly came much later. By studying photographs such as this, scientists have been able to foster theories of Mars' early history. *Right:* The Orbiter Performance Analysis Group during the landing of Viking II.

place. Other gases detected include 1.5 parts per million of xenon, .9 parts per million of krypton, and 10 parts per million of neon.

[Characteristically, as soon as this data was received, scientists began disputing what it meant. Dr. Michael McElroy felt that the evidence indicates that Mars most certainly had a thick atmosphere in its youth. Dr. Tobias Owen, on the other hand, said that the information was much too sketchy to suggest any such conclusion! Clearly, the setting was ripe for the life experiments. However, before we turn to this unprecedented search, as well as other Viking I findings, let's pick up the trail of Viking II. Together, the two crafts will provide a rounded picture of our current knowledge about Mars.

[Viking II achieved Mars orbit on August 7, 1976, and a bold program was planned for the lander's descent. The craft was to set down somewhere in a 10-degree swath halfway between the north pole and the equator. However, the site would be chosen for its scientific interest rather than its safety. Mission directors were not being reckless; but they were willing to gamble with this second craft in the hope of gaining greater scientific rewards. And the site they decided upon was a plain named Utopia, a spot 4,000 miles northeast of where Viking I had landed. The reasons for the selection were outlined by Harold Masursky of the landing team. "The Viking I landing site took place at a latitude about equivalent to Yucatan or Mexico City, whereas the second landing is supposed to take place at

Above: The first photograph taken by Viking II after its landing on Mars. Rocks in the 4- to 8-inch range are visible; the small boulder just above the lander's footpad displays a dust-covered or scraped surface, suggesting that it was overturned or changed by the landing maneuvers. The center of the picture is about 5 feet from the camera. *Below:* The video-oriented Spacecraft Operations Test Area.

a latitude equivalent to Montreal. So we have an enormous spread in latitude and the climate on the earth would be very different in the two areas. So our intention was to land in two regions as different as possible. For, in terms of climate and the water content, and the temperature, this is certainly the case. The materials covering the surface, we think are uniform dune material. They may look somewhat like the small scatter dunes that we landed at the edge of in the Viking I site. It may be that since these have a much better dune shape, that the materials composing them might be coarser grained. That is, they might be medium sand size rather than the very fine materials in the sampled area. We anticipate that the area won't be free of blocks because there are a number of fresh craters in the area that will undoubtedly have punctured through that thickness of dune material, and that will bring up what we think are the subjacent volcanic materials and spread them out on the surface and hopefully there'll be small fragments that will be within reach of the lander. We think we should be able to get adequate samples in the site, and since they're spread considerably in latitude and we have a high water content (as determined by orbiter readings), hopefully the conditions that the Gas Chromatograph/Mass Spectrometer and biology samples we'll work on will be considerably different from the first site." Thus, at 12:40 in the afternoon of September 3, 1976, the Viking II lander began its controlled fall to the surface of Mars. It arrived safely on the planet at 3:58:20 P.M., PDT. Yet, for several hours, the feeling at JPL had been that the mission might be over before it had begun.

[Shortly after separation, technicians lost all contact with their two probes. Apparently, the jolt of disengagement had caused the orbiter to rock and lose sight of its orientation star Vega. With its high-gain antenna no longer aligned with earth, we could receive only weak signals from its low-gain unit. The real fear, however, was not so much in the landing maneuvers—Viking II, like its predecessor, was on a preprogramed sequence. Rather, scientists were worried that the data gathered by the lander and set for automatic relay to earth through the orbiter, would be inexorably lost in space. Thus, they ordered the beleaguered orbiter (by addressing its low gain antenna) to record instead of broadcast the information. The only message they allowed it to send was the news that the lander was safely settled on the Martian surface. Receiving this faint transmission, Project Manager James Martin said, "I'm confident we're down

Above: Two early Viking II photographs created this mosaic of the lander's northeast view, stretching two miles to the horizon. A small channel of some sort crosses from the mid-left to the lower right of the photograph. *Below:* The Viking lander model at the Jet Propulsion Laboratory, here tilted 8.2 degrees to correspond with the slant of the Viking II lander on Mars.

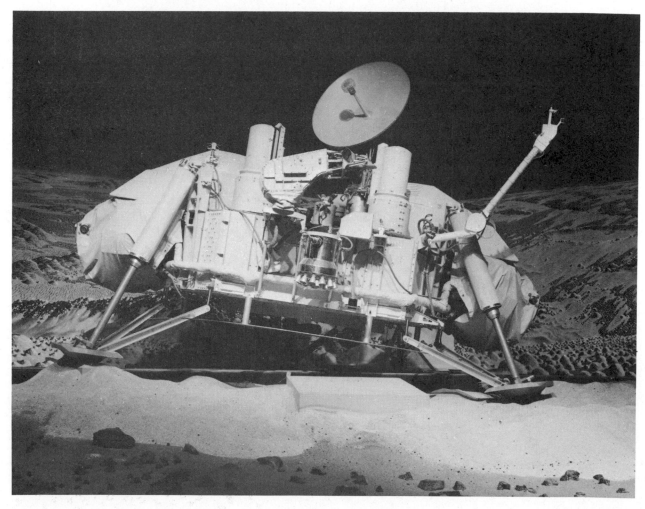

safely." He added, "That lander doesn't seem to need us." And, while the Viking II descent lab took a few pictures and got used to its new environment, scientists had the orbiter perform a Mariner IV-ish maneuver and roll about as it circled Mars until it was able to find and lock on Vega. The recorded data and photographs were beamed earthward shortly thereafter.

[The initial communications of Viking II showed that the lander had set down with one footpad on a small boulder, and was tilted 8.2 degrees to the west. Dr. Gerald Soffen, Chief Project Scientist for Viking, remarked, "Maybe it's a Martian that's got it propped up." In any case, the angle presented no problem other than that photographs of the horizon were distorted due to the incompatability of the camera's axis with the lander's position. Computer reprocessing took the differentiation into consideration and restored such pictures to their true perspective. As for the weather, local temperatures were not much different from those at Chryse, ranging from −128 degrees to −35 degrees F. The winds were blowing at an average speed of 30 miles per hour. Other analyses, including pressure (7.7 millibars) roughly approximated those of Viking I. The only real surprise was that, as imaging scientist Dr. Thomas Mutch put it, Viking had landed in what "is certainly not a well-winnowed sand dune—that's for super-sure!" Instead, the craft was secluded on a flat field of small rocks. Dr. Elliott Morris of the U.S. Geological Survey described them based on imaging data during a September 9 press conference (see accompanying photographs to observe the features described by Morris). During his talk, Dr. Morris also gave the audience some insight into the agony that is unique to the inquisitive mind of the scientist. "The surface is littered with all kinds of blocks, which are very exciting to us, as geologists. I sometimes, in looking at these pictures, get very frustrated. It's something like a little boy outside a bakery shop with his face against the window, seeing all these goodies and smelling it—and can't go in and sample it. We'd like to get out there and get ahold of these blocks and break 'em up and look at them under a hand lens to see what they actually are. All we can do is look at 'em and speculate on what we think we see. Some of the kinds of the things that we see is here we have these very porous, sponge-looking rocks—vesicular rocks, we call them. And they range from these porous varieties to very fine-grain types of rocks. The porous ones range from quite dark to quite light. The fine grain rocks appear to be quite light. They are a variety of shapes. Most of them

173

Above top: Viking II's second photograph of Mars' surface. The bend of the horizon in the east is an illusion caused by the lander's 8.2-degree tilt toward the west. *Above bottom:* When reprocessed by computer, Viking II's second photograph is quite improved. It covers all but 36 degrees of the lander's full panoramic view. As for the rocks, many of the pitted samples resemble lava, while others may have been fluted by the wind. There is no indication of the sand dunes that mission leaders had hoped to find! *Below:* The southern view of Viking II.

are quite angular. We see a few rounded ones. Also, between the rocks the surface is fairly coarse compared to what we saw in the inner block areas of the first landing site. A number of smaller grains have a pebbly surface. And then on top of that, we see these fine grain drifts, similar to the fine grain material we saw at the first site. And these drifts are between the blocks. In some places we see the drifts appear to be banked on the southeast sides of the blocks. And sometimes in the back of these blocks we see the fine grain material kind of banked up in little banks. We don't see the nice wind tails as we saw on the first site. They're there, but they're not as well developed. Other things that are interesting—as we look way out at the far horizon, we see some flat-topped hills. These hills are quite bright in the afternoon sun. We can't tell yet whether or not these are actually craters or whether they're ridges or just flat-topped hills." Tellingly, Dr. Morris concluded by saying that he felt "right at home" looking at many of the rocks, indicating that a great majority of them are akin to formations found on earth. This further substantiates the common origins theory of the planets as described in chapter two.

[Beyond the different terrain, the one immediate advantage which the second lander had over the first was that this time, the seismometer dutifully uncaged. Dr. Gary Latham reported in a September 5 news briefing on the instrument's activities, "We received the data this morning covering the first 20 hours of operation or so. From this data we were able to see that the seismometer did uncage successfully about two hours and 40 minutes after touchdown. All of the functions that are supposed to be performed during these initial days of operations work performed as expected, and we see no evidence suggesting any problem whatsoever. One very pleasant surprise is that at least the preliminary data suggests that the Mars environment will be quieter than we might have anticipated; by quiet, I mean in terms of surface vibration. We believe now that we can use the full sensitivity of our instrument. Many of us were somewhat concerned that wind effects, for example, on that lander would have caused vibration which would have made it difficult for us to record true ground vibration. That is not happening. If we can indeed operate at these sensitivities, we can magnify the surface motions of the planet about 200 thousand times. This will be equivalent to a quite good seismometer on the surface of the earth. In that event, we should be able to see something equivalent say to a magnitude-six earthquake

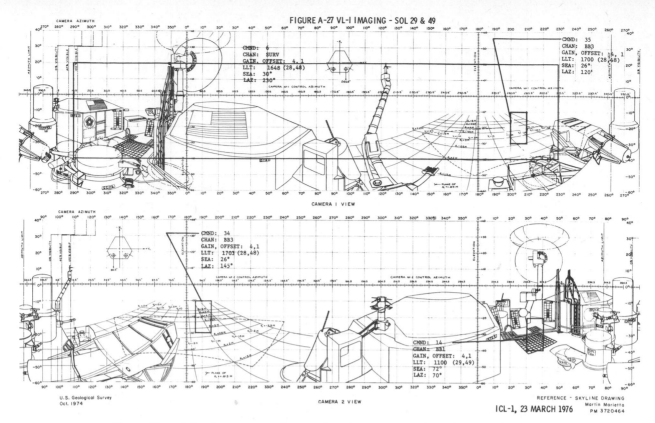

U.S. Geological Survey
Oct. 1974

CAMERA 2 VIEW

REFERENCE - SKYLINE DRAWING
Martin Marietta

ICL-1, 23 MARCH 1976   PM 3720464

Above: One of the dozens upon dozens of grids used to analyze photographs received against the possible range of the Viking cameras. This pair of grids represent the scope of Viking I. *Below:* White circles indicate the spot where the Viking II sample arm ultimately dug its first trenches on Mars. The photo on the right was taken by camera #2; the photo on the left by camera #1. Together, the pictures provide a stereoscopic view of the terrain. The computer-generated grid lines form a map of the area. Comparing this map with orders stored in the lander's computer, it was learned that the scoop's pre-programing would have sent it burrowing right into a rock. Thus, it was told to dig into the circled area instead. Notice that the vertical lines are all relief interpretations of the terrain; the intersecting line represents the reach of the sampler arm.

originating anywhere on the planet. There are something like 10 or 20 of those sized events on earth every year, so we would have a good chance if Mars is as active as the earth." Finishing his rundown, Dr. Latham added wryly, "And of course, it's a pleasure to join the working experiment club after waiting so long!" As for his findings, by the middle of October 1976, no Mars quakes had been detected. However, most scientists are convinced that they do occur. As Dr. Don Anderson of the California Institute of Technology observed, based on orbiter photographic reconnaissance, "The younger volcanic terrains of Mars . . . are not just icebergs sitting in a fluid, but they are exerting stresses on the crust. If you calculate the stresses that are involved by this lack of compensation, then these are extraordinarily large stresses. Therefore, I would expect there to be releases of these stresses in the crust, and I would expect large Mars quakes."

[Several weeks of general research conducted by the crafts gave us a more rounded picture of Mars, although there were few upsets of previous findings. The one major embellishment of earlier readings centered on the change in local pressures. Viking II landed .78 miles lower than its companion craft. Yet, scientists found that the pressure at both sites, inconsiderate of depth or location, was dropping. This indicated that carbon dioxide was, indeed, being drawn from the atmosphere and frozen about the poles. Thus, in addition to the gas losses which had been detected by Mariner IX, and the presence of argon noted earlier in the mission, scientists now had a comparatively sophisticated portrait of the history and chemistry of the Martian atmosphere. However, from eight days after the landing of Viking I, the greatest attention in both the media and most scientific circles was given to the trio of biology experiments. There were some problems, at first, with the sampler arm of the first lander. Going through its test routine, the arm extended six inches but was unable to retract. An uneasy group of Viking engineers duplicated the situation on the life-size model of the craft at the JPL and found what appeared to be the problem: a three-inch-long locking pin had failed to dislodge. The arm on the prototype was extended and the pin popped out; the Mars craft was told to duplicate the move, which was successful. The sampler stalled again on August 3, but was repaired within four days. The experiments were conducted without any further mechanical mishaps. The Viking II life search began on September 13, although, once more, the soil sampler caused some consternation. The collector

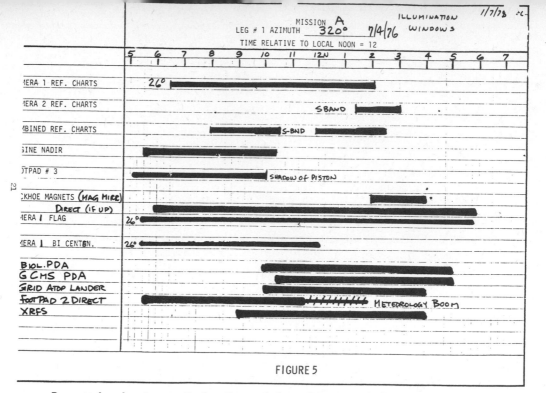

MISSION A
LEG # 1 AZIMUTH 320° 7/4/76   ILLUMINATION WINDOWS
TIME RELATIVE TO LOCAL NOON = 12

FIGURE 5

Bar graphs showing particular times of day when spacecraft components are un-shadowed. This diagram is from the Viking I lander.

head which shovels up the Martian soil jammed. At first it was thought that a rock had become lodged in the mechanism—but by September 15, the problem was diagnosed as a faulty switch and was overcome by a few commands from earth.

[Before we look at the individual life experiments, we had best provide ourselves with a general guideline by summing up their results. In a word, they were inconclusive. Although, as of this writing, the two crafts are still operative and poking around for Martian biota, there is little chance that they will make a definite determination about life on Mars one way or the other. Indeed, there has been so much information and so many theories advanced since the Viking landings that, if anything, the waters have tended to get more muddy with each passing day. As Soffen put it, he had always expected the robots to "flick clues at us rather than bombard us with answers."

[What kind of "clues" have the landers sent us? Let's take a look at the Pyrolytic Release Experiment, which is typical of the confusion fostered by Viking. The question asked by the test: does the Martian soil process carbon dioxide and carbon monoxide the way a living sample of earth soil would? Reduced to its simplest determination, the higher the number recorded by the experiment, the more likely the existence of life processes. The first time the experiment was run, on Viking I, the number was a huge 96 . . . and biologists were thrilled.

The operation of the surface sampler is illustrated in this series of photographs. This small area, christened the "Bonneville Salt Flats," offered unique opportunities for organic scavenging due to its thin crust in relation to surrounding materials. There were, of course, some dangers—such as the 8-inch-high rock in the foreground. However, in the first shot, the arm has negotiated past the rock and the backhoe has penetrated the surface about one-half inch. In the second picture the scoop has been pulled back to avoid a collision between a rock in the shadow of the arm and a plate joining the arm and the scoop. In the third picture, the scoop has been thrust forward, acquiring its sample; retracted, the arm reveals the trench it has created, a hole 3 inches wide and some 2 inches deep.

Soffen did not hesitate to state, "If we had found this reaction on earth, where we know there are organic compounds, we would have no hesitation in saying that there is biology at work." But no one was ready to declare Mars a living world. The test was performed a second time, with one slight difference. On this run, the sample was heated to a temperature that would have killed any comparable terrestrial life forms. The rationale for this imagined slaughter was that if the number remained 96, then the reaction had to be of a chemical rather than a biological nature. The results were too good to be true. The number was a sluggish 15. Scientists felt that they might have found life. However, just to be sure, they repeated the first experiment—and were in for a shock. The number was a monkey-wrenching 28. Viking II performed the Pyrolytic Release, although without the xenon lamp, and came up with a figure of 21. The thinking was that whatever had processed the carbon dioxide and carbon monoxide was active only in the light. But how to explain that 28? And, compounding the problem was the fact that the Gas Chromatograph test had failed to turn up any trace of organic chemistry in the soil. Without such activity, of course, there can be no life.

[The other life-search experiments offered scientists little solace. At the July 31 news conference, Dr. Harold Klein, leader of the Biology Team, described the initial results from the Gas Exchange Experiment and the Labeled Release Experiment as follows: "What we have found

179

out basically comes down to two things, two important and unique and exciting things. In one of our experiments, the Gas Exchange Experiment, we believe that we have at least preliminary evidence for a very active surface material." What had happened was this: as soon as the nutrient was introduced, a greal deal of oxygen was released into the atmosphere, over 15 times the amount for which scientists could account. If it is not life, then the reaction could have been caused by the fact that, as a desert, the landing site has not experienced humidity for quite some time. The introduction of the liquid may simply have triggered some kind of oxidizing reaction in the iron oxide coating of the surface particles. "We believe," Klein suggested, "that there's something in the surface, some chemical or physical entity which is affording the surface material a great activity. And may, in fact, mimic—and let me emphasize that—may *mimic*, in some respects, biological activity." On the Labeled Release Experiment, Klein noted that, "the preliminary data indicates that the test cell is elaborating a fairly high level of radioactivity which, to a first approximation, would look very much like a biological signal." The scenario for this experiment saw the liquid nutrients injected and, shortly thereafter, radioactive carbon was found in the atmosphere. Scientists were careful to offer a nonbiological explanation, lest the discoveries be misinterpreted: like the Gas Exchange Experiment, activity may have been stimulated in the oxidized soil coating, caus-

The first panorama of the Martian surface taken by Viking I. The view covers 300 degrees. Portions of the lander are visible in the picture, including the sample arm on the left, the low gain antenna in the center, color charts for camera calibration and the magnetic properties experiment mirror and the high gain antenna on the right.

ing the disintegration of the labeled nutrients and thus releasing carbon-14. Klein summed up these findings by observing that, "combined with our other information on the Gas Exchange Experiment (it is apparent that) the soil is extremely active. These results must be viewed very, very carefully in order to be certain that we are, in fact, dealing with a biological or nonbiological phenomenon. If it *is* a biological response, it is stronger than anything we have obtained from terrestrial soil. It would mean that biology on Mars is more highly developed, more intense, than life on earth." Dr. Norman Horowitz of the Biology Team added his own thoughts on the matter: "If it turns out that [these] results can be explained by inorganic chemistry, this is a chemistry that certainly is not occurring on earth." Dr. Gilbert Levin, also of Biology, agreed. "All we can say, at this point, is that the response is very interesting. Be it biological or nonbiological, it's unanticipated." Dr. Levin went on to address a concern which had permeated the entire Mars program: "The possibility that we may have taken an earth organism with us to Mars, and are now detecting it or its progeny, has been considered. There was much ado about planetary quarantine in the Viking project, and we are assured that there is less than one chance in a million that we carried this organism to Mars with us. And, furthermore, the amplitude and kinetics of the response would rule out the possibility that the organisms, if there are any organisms there, came from earth.

181

Left: This Viking I photograph covers a field of 13 feet at the foot of the picture, to 1.8 miles at the horizon. The photograph shows an accumulation of fine-grained material behind the rocks, which suggests wind deposition of dust and sand downwind of the blocks. The distance across the horizon is 110 feet, and the boulder at the top of the photograph is 13 feet wide. *Right:* Another photograph for the alien-hunters among you! The rock on the far left of the photograph appears to have a "B" etched in its side. The result of weathering, or a message from chariot-riding gods?

I think, in conclusion, we must say that we cannot reach any conclusion about whether or not this is a biological response."

[By August 20, the interpretation of data was relatively unchanged. Of course, Viking II had not yet set down, but Dr. Klein summarized the experiments to date by saying, "I know this must seem to be a badminton game between chemistry and biology. At any rate . . . biology seems at least to have survived, if not to have gained some advantage over chemical explanations." On September 14, he had this update, which was less a scientific analysis than a progress report: "On Viking I, we are continuing the Gas Exchange incubation and the Labeled Release incubation, at the present time. The Gas Exchange is showing a slight increase in carbon dioxide each time we have made a measurement. The Labeled Release Experiment in the first lander is still continuing its third analysis. At the beginning of that third analysis, we had a rapid release of radioactive gas approximately equivalent to the data that we saw the very first time. That experiment has continued to incubate now and still is showing a slight increase in radioactive gas each day. Pyrolytic Release Experiment is quiet at the moment on lander I. Now, focusing on lander II, we're only into the first day's worth of data. That's all I can say about that one." Two days later, Dr. Klein reported, "In the case of the

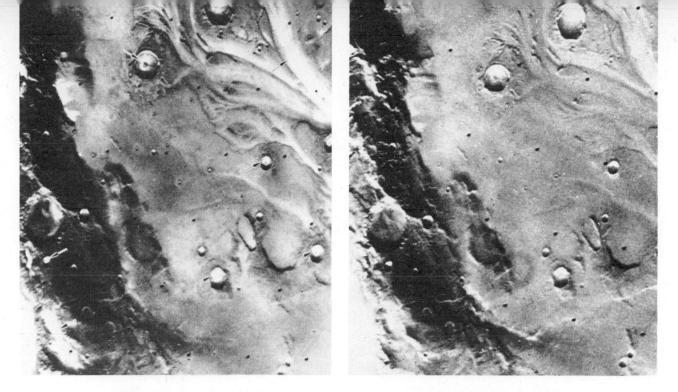

Two pictures taken a half hour apart by the Viking I orbiter show the formation of early morning fog in low-lying regions of Mars. The picture on the left shows water vapor as indicated by the arrows; thirty minutes later, a slight warming caused by the rising of the sun has driven off the mist.

second landing site, the preliminary data indicate that we are in a region which is responding qualitatively very similarly to what we saw at the first landing site." Then, on September 22, a press conference was called to discuss a discovery that is crucial to the presence or absence of life on Mars: the identification of water as the major component of the polar caps. In September of 1977, nearly one year later, large patches of frost began to collect at the Viking II site, which is less than 43 degrees of latitude from Mars' North Pole. According to Dr. Gerald Soffen, this meant that while, "I don't expect we're going to find any massive amounts of liquid water today . . . there could easily have been flowing water in the past history of Mars." However, Dr. C. Barney Farmer, leader of the Water Vapor Mapping Team, added a disconcerting footnote about the unavailability of this water to present biota: "It seems that what happens is that as the northern cap passes into the winter, which is what it's just about to start now, and the temperatures fall and the vapor that's in that region rapidly condenses back onto the surface, the first thing that we grow is a surface-extended cap. As the temperatures fall even further, then they become low enough for the carbon dioxide to start to condense at that end of the planet on *top* of the water, and, of course, any water that's still around is trapped. [Although] the reverse happens in the follow-

183

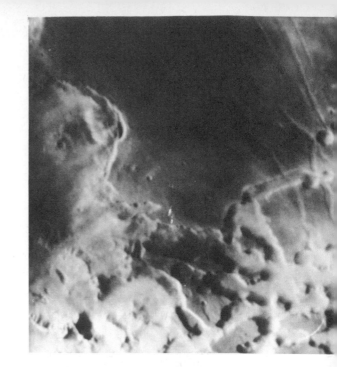

Above left: These overlapping photographs taken by the Viking II orbiter shows the melting, mid-summer north polar cap. *Above right:* As the sun rises over Noctis Labyrinthus, bright clouds of water ice can be seen about the tributary canyons of this high plateau region of Mars. *Below:* The incredible west end of the Valles Marineris as seen from 2,700 miles up. These two particular canyons are the Tithonium Chasma *(top)* and the Ius Chasma. They are each approximately 37 miles wide and nearly .7 miles deep. They were formed by the downfaulting of the crust along parallel faults. Erosion whittled the canyons to their current state, in which impact craters can also be seen.

Above: Over 100 individual photographs from the photomosaic of the Valles Marineris. This is a broad overview of the photograph showing Tithonium Chasma and Ius Chasma. Scientists are relatively certain of the processes that created the valleys, but are hard-pressed to explain where the material which once filled them has gone! *Below:* Phobos as viewed by the Viking I orbiter. This is a heavily cratered side of the Martian moon that was not visible in the Mariner IX photograph.

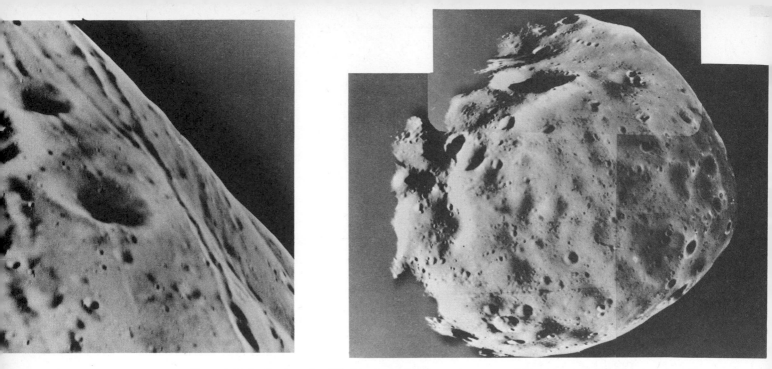

Above left: A photograph of Phobos from the closest range at which any spacecraft has ever photographed the small satellite. The range in this February 20, 1977, photograph was 75 miles. The area covered in the Viking I orbiter picture is 1.86 by 2.17 miles. *Above right:* This incredible mosaic of Phobos was taken from 300 miles away by the Viking I orbiter on February 18, 1977. The image of Phobos covers 13 by 11.8 miles. *Below:* A phenomenally detailed photograph of Phobos taken by the Viking II orbiter. This picture was recorded from a comparatively close range of 545 miles.

ing spring and summer, it does appear that the annual budget of water vapor does not allow for a very large percentage of that residual water ice at the cap to become part of the atmospheric inventory." Thus, while an intriguing and important discovery, the meaning to the Martian environment of water ice caps is something that will become plain only after many, many months—perhaps years—of study.

[The next important biological maneuver came on October 4, when the Viking scoop began overturning Martian rocks to dig for soil samples. The reason for this, explained Dr. Henry Moore of the U.S. Geological Survey, is "to try to get a sample that has escaped the ultraviolet radiation on Mars, and also . . . if the surface materials are related to windblown dust, then surely a sample from beneath a rock, which wouldn't move in a wind, would not be contaminated with such windblown dust, which might be planetwide." Unfortunately, as with the samples taken from exposed areas of the Martian surface, the results of analysis on the scoopfuls gathered by both sample arms were tantalizing . . . but inconclusive. Even as studies continued late into 1977 and early 1978, the question of where organic materials could be hiding on Mars plagued scientists. For if life did not exist on the Red Planet, the scarcity of organic materials in the face of water and carbon-rich meteorites which once pelted the planet is mystifying. One exciting theory suggests that cannibalistic Martian microorganisms might travel the planet and consume the remains of such matter, these tribes being too small or too localized to be detected by Viking. Another hypothesis, put forth by A. A. Mills of the University of Leicester, England, proposes that during the planet's awesome dust storms, friction within the dust clouds may build up electrostatic charges that, when they discharge, would be, as Mills puts it, "a very effective scavenger of even the most tenacious traces of organic matter . . ." Such discharges may also create ozone from molecular gases, which could cause the oxidation that many scientists consider to be a major contributor to Mars' reddish hue. All of which reminds one of the popular scientific anecdote about a newspaper editor who, late in the fifties, fired a telegram to a leading astronomer. "Doing feature on space. Please cable one hundred words on question Is There Life on Mars?" The scientist replied promptly, wiring, "Nobody Knows," 50 times.

[Even after landing two versatile robots on the Red Planet, the answer to the question Is There Life on Mars is still "Nobody Knows." The happy difference is that now, at least, we have more facts at our disposal than ever before, and will surely press on in this remarkable investigation. The question, of course, is what form will future research take? Let's turn first to the unmanned probes. There will be other Vikings and, indeed, Viking III is already on the drawing boards for launch in the 1980s. In September of 1977, a typically and frustratingly nearsighted Congress shot down pre-construction development of the probe but, hopefully, NASA will prevail in an upcoming session. Although there will be refinements in the vehicle's instruments based on what are bound to be interpreted as weaknesses in Vikings I and II, the greatest improvement will be the addition of wheels to the craft. It will be able to travel for several miles at a stretch, stop, and take a photograph of the upcoming terrain, beam the picture to earth, and wait for us to chart and program a course. It will even be able to compare the red to red-less sections of Mars to see if organic materials may be plentiful away from the electrostatic discharges described by A.A. Mills. Needless to say, such a vehicle will be an invaluable surrogate explorer for all the scientific teams. In terms of other mechanical astronauts, the Soviets have landed a photo-taking probe on Venus (they found a rocky landscape not unlike that of Mars), although our only coming exploration of that planet is a Mariner craft to be launched in 1978. There are plans to land a Viking-like ship on Titan, an astonishingly vital moon of Saturn and the only world beside earth and Mars which is seen as a possible harbinger of life. There is also a proposed robot voyage to Halley's Comet, and the asteroid shot mentioned in chapter five. Indeed, the asteroid mission has gained impetus from Viking studies of Phobos and Deimos. Dr. Joseph Veverka of the Orbiter Imaging Team summarized the Viking studies:

For the first time we have adequate resolution in the images of both satellites to be able to say what the density of the craters is on the surfaces, and that tells us something about the age of the surfaces. The problem with the Mariner IX imagery was that the resolution was too poor to see an adequate number of craters. Deimos is completely saturated with craters. Now, what does that mean? If you expose a surface to meteroids out in space,

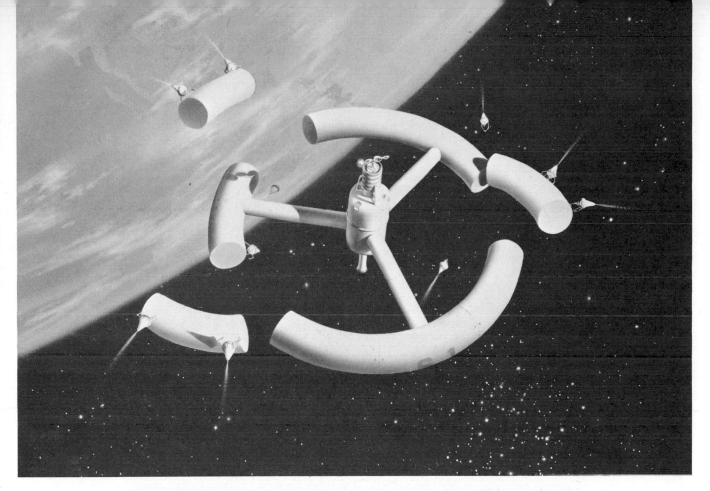

Above: Walt Disney Productions' idea of a space station from which interplanetary vehicles can be launched. *(From the television special* Man and the Moon, *copyright by Walt Disney Productions).* Below: According to the original caption, "Crew members of a Martian ship observe on a television screen, the progress of the line of other ships in the first expedition to the planet Mars in this scene from *Mars and Beyond*." *(Copyright by Walt Disney Productions).*

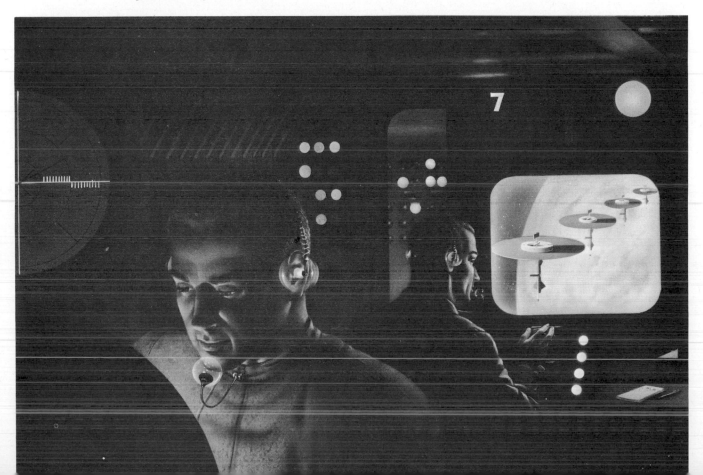

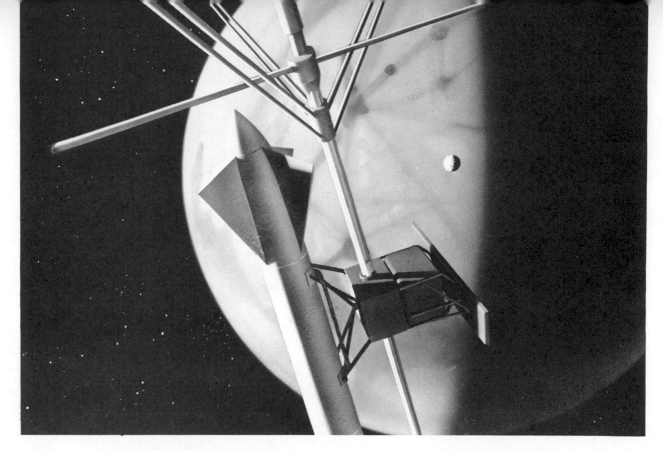

Above: As a Martian ship comes to within 4,000 miles of Mars, they get a closeup view of Phobos in Walt Disney's *Mars and Beyond. (Copyright by Walt Disney Productions).*
Below: A pressurized, domed community suggested for Mars by Walt Disney in his television presentation *Mars and Beyond. (Copyright by Walt Disney Productions).*

These stamps, commemorating the Viking landings, were issued by the African Republic of Togo, Dominica, and Grenada in the Caribbean.

there is a maximum surface density of craters that you can get on the surface. Because after a while the processes which produce craters and those which obliterate them are an equilibrium and you're knocking out as many craters as you're producing. So there is a limiting density on the surface of any object that you could produce. This saturation density is reached, for example, in the most heavily cratered regions of the lunar uplands. From similar images of Phobos, we have also concluded that the surface of Phobos is also saturated with craters. Now to be able to translate 'saturated' into an age, you have to be able to say something about what the flux—or the rate at which craters are being produced. That number at the orbit of Mars is not very well known, but a very conservative estimate of that would say that this means the surfaces are at least two billion years old and are probably a lot older than that. There are also many things that we are seeing in these images which, at the present time, we don't understand [like] the famous grooves on Phobos. Let me give you one hypothesis, which is especially exciting. There parallel grooves may actually represent the layering in the surface of a large parent body of which Phobos is a small fragment. Now if that were the case, then this layering should run clear across Phobos. And we should see expressions of this on the other side. So that is one reason why we're very anxious to get images of this resolution of the other side of Phobos to see whether this structure is repeated there. But until we do that, we really can't say anything very definitive about the origin of these. As for color, Phobos and Deimos are perfectly gray. Now this tells us something about the composition of the material of the surface, because we know that the material is very dark; it's very gray,

191

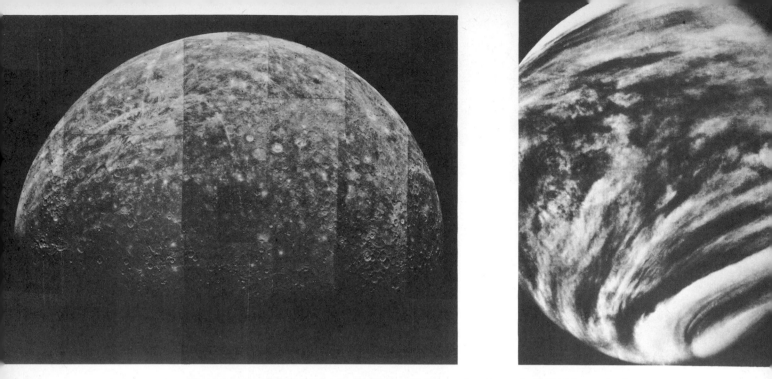

Above left: In NASA's future are doubtless further explorations of all the planets in the solar system. This photomosaic of Mercury was constructed from shots taken by Mariner X on March 29, 1974, when the vessel was 130,000 miles from the planet. *Above right:* A Mariner X photograph of Venus, taken on February 6, 1974, from 450,000 miles away. *Below:* Developed by NASA's Marshall Space Flight Center in Huntsville, Alabama, this Elastic Loop Mobility System will no doubt allow such future vehicles as Viking III to roam about the surface of Mars. *(Illustration by C. Bennett).*

and apparently, if you changed the particle size of it, it doesn't change color very much. And that suggests that there's something very opaque in the surface and one possibility is carbon. And one consistent picture is to say that both Phobos and Deimos may consist of carbonaceous chondrite material—material which is very common in certain meteorites that fall to earth. Now, the reason this is so very important is that there is only one place that you can really make this sort of material, when you're forming planets five billion years ago, and that is at about three astronomical units (a measurement based on the distance from the earth to the sun, or approximately 92,900,000 miles) in the asteroid world. So if it does turn out that we can establish that both Phobos and Deimos are made of carbonaceous chondrite material, then they are almost certainly pieces of something that formed in the asteroid belt and they have been captured by Mars.

[Understandably, these missions and others like them hold little fascination for the general public. Where, then, do we stand on the more glamorous matter of firsthand human exploration of Mars and the other planets? Many scientists are against it, opposed to the political grandstanding and comparative expense that manned flights have thus far represented in space. Yet, it is, paradoxically, the kind of space exploration which can most successfuly grab the public's imagination.

[Historically, science and science fiction writers have envisioned primarily manned voyages to Mars. In fact, the only prominent mention of a robot probe in a popular science text was made by Arthur C. Clarke (author of *2001: A Space Odyssey*) in his 1959 classic *The Exploration of Space*. Clarke envisioned a craft that would look "peculiar to anyone who imagines that rockets must be sleek, streamlined projectiles with sharply pointed noses," observing that "such refinements are not only unnecessary but actually wasteful on rockets which are launched in airless space." Clarke predicted, and accurately so, that rockets would only be used to carry these probes into space, after which they would unfold "outrigger arms" and be on their way. Arriving at Mars (the instance selected by Clarke), they would be run by small electronic brains and told to photograph the planet. He did not see live television as feasible across these great distances, but suggested that still pictures might be sent earthward every few

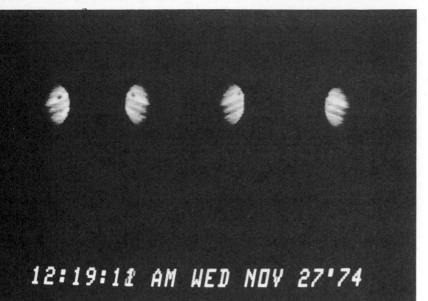

12:19:18 AM WED NOV 27'74

NASA   PIONEER 11   UN
RANGE: 3400000 KM   PHASE: 46 LCM2
MID TIME OF DATA RECEIPT.   30 NOV 17
CS4   BLUE   DATE   12

Above left: Pioneer XI's photographs of Jupiter from a distance of 4 million miles. *Above right:* Photographs of Jupiter by Pioneer XI while it was just over 2 million miles from the planet. *Below:* A proposed lunar rover, fully automated, which would gather soil as well as try to produce water and oxygen.

minutes, riding on radio beams. In his only understatement of this remarkably prophetic volume, Clarke stated that these photographs would prove "quite adequate" in the realm of planetary exploration. However, in *The Exploration of Space*, Clarke was writing with the benefit of several successful earth satellites behind him. This is not to derrogate Clarke's speculative work; rather, it is to laud a trail-blazing monograph entitled *Mars Project* and written by Dr. Wernher von Braun in 1952. At the time, composing a detailed program for a manned expedition to Mars was the playground of the fantasist—not the respected scientist. Which is precisely the reason that von Braun wrote the treatise. He was tired of science fiction films and books in which "the central figure was usually the heroic inventor. Surrounded by a little band of faithful followers, he secretly built a mysteriously streamlined space vessel in a remote backyard. Then, at the hour of midnight, he and his crew soared into the solar system to brave untold perils. [And] while a man encased in an impressively clumsy pressure suit, walking importantly around the base of a space vessel, makes a fine and interesting figure in a lunar or Martian moving picture,[17] it is unlikely that he will gather much useful data about the heavenly body on which he stands." Von Braun had a few more practical ideas, which he outlined in impressive detail. And, although this is probably not the way that we will eventually travel to Mars, one has the gut feeling that this is the way it *should* have been!

The rocket scientist's Martian expedition consisted of a flotilla of 10 vessels, each one manned by not less than 70 people.[18] Each ship of this space navy would be assembled in a two-hour-long earth-orbital path, to which three-stage shuttle rockets would ferry all structural material, propellants, provisions, and personnel. Being assembled beyond the tug of earth's gravity would, of course, allow the ships to fly into an elliptical, heliocentric orbit with minimal thrust. This course, tangent with the orbit of Mars, would bring the planet and the flotilla together several months later, at which time orbital insertion maneuvers would be executed. When all the crewmembers had performed their specific duties and observations, descent to the surface

17 Dr. von Braun was referring, in this instance, to both *Destination Moon* and *Rocketship XM*.
18 Von Braun defends his flotilla-to-Mars idea by noting that other pioneers, like Columbus, never traveled alone. This, of course, was in the days before Congress and their various budgetary indiscretions.

Above: A proposed Venus radar orbiter: while no camera could possibly penetrate the planet's thick clouds, a satellite in a 625-mile-high orbit could topographically scan the entire globe, piercing the clouds via radar. The orbiter might also drop an instrument package to the surface of Venus. *Below:* A proposed mid-1980s robot exploration of the Asteroid Belt. To quote the original caption, "It would be a chance to go rummaging, so to speak, in the cosmic garbage."

Above: A mission to the remarkable earth-like moon Titan, in orbit around Saturn, is something that scientists are working hard to get on the NASA schedule. Dropping a laboratory onto the satellite, the parent craft would remain in orbit to study both Titan and Saturn. *Below:* A hoped-for exploration of Mars in 1984. Two pairs of rovers would scour the Martian surface at the rate of 3 miles per day, covering a wide range of missions or helping one another should the situation arise.

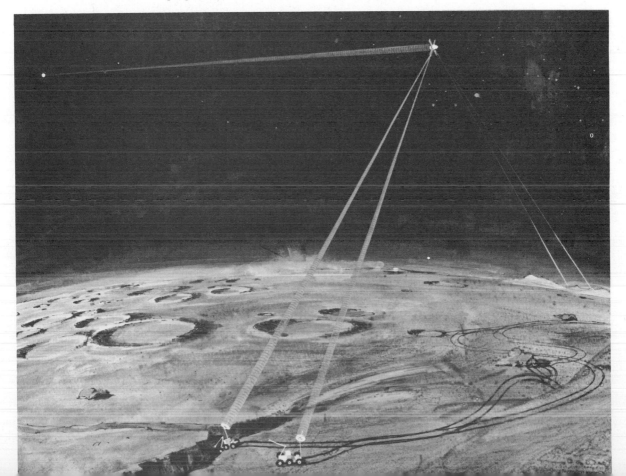

Above: An interesting Ken Hodges rendering of a pair of Martian rovers. *Below:* Late in the 1980s, scientists plan to have rovers gather samples of Martian soil and rock, return them to a waiting shuttle craft, which would then carry the booty to a solar sail propelled interplanetary ship. Solar sailing, the use of ionized protons to go tacking about the solar system, is economical and the most likely means of travel in the future.

would begin. However, this was not to be the science fiction writer's dream of a rocket backing vertically to the surface on a pillar of flame. Von Braun thought to equip three of his 10 rockets with landing boats, and here is how the scientist saw the landing:

> Considering the risk attending a wheel landing on completely strange territory at relatively high speeds, it is assumed that the first landing boat will make contact with the Martian surface on a snow-covered polar area, and on skis or runners, minimizing the risk. This boat will be abandoned on Mars because of the impossibility of reascending from polar latitudes to an orbit in the plane of the elliptic. So this boat need not carry any fuel for the re-ascension. The total useful load will be 125 tons, part of which will consist of ground vehicles. With such vehicles, the crew of the first landing boat would proceed to the Martian equator and there select or prepare a suitable strip for the wheeled landing gears of the remaining two boats. At the termination of work on Mars, the crew of the first boat would return in the remaining two boats to the waiting passenger ships.

[The landing crew would then be transshipped to the seven remaining vessels, the three which bore the descent boats being abandoned in Martian orbit. The ships return home and, from earth orbit, the same shuttle crafts which had helped build the interplanetary vehicles retrieve all personnel, equipment, and Martian finds.

[In terms of specifics, von Braun saw the total expedition lasting two years and 239 days. Fifty people would spend a total of 400 days on the planet, during which time 950 ferry flights would be required between the surface and the flotilla. The predicted cost of propellants for the expedition (seen by von Braun as the major expense of the project) was $500 million, most of which went to the Mars-orbit shuttle flights.[19] Other details accounted for by von Braun in his careful study include projected skin temperature of the ferries, their wing area, the volumetric content of their fuel tanks, the trajectory of the ships, and

19 As for particulars on the fuels themselves, von Braun said, "The use of liquefied gases as propellants is purposely avoided. In theory, liquid hydrogen and oxygen would save a great deal of weight, but their low boiling points would complicate the problem. Thus, all computations for ferry vessels, interplanetary ships, and landing craft, are based on hydrazine and nitric acid. These propellants remain in liquid phase at normal temperature."

so forth. He even computed the weight of each vessel, from the radio equipment (one metric ton)[20] to ten crewmembers (.8 metric ton) to personal baggage (1.2 metric tons) to food wrappings (one metric ton) to space suits (.5 metric ton).

[While von Braun's plan may not have set a massive Mars program into action, it *was* one of the works which inspired a literate look at the Red Planet by Walt Disney Productions. Shown during the 1957–58 season of the *Disneyland* television series, *Mars and Beyond* was the first mass-media, documentary look at our celestial neighbor. According to press releases, the purpose of the show was to measure what we knew about Mars against what we surmised about the planet. "When earthman finally walks upon the sands of Mars," one of the dozen promotional sheets began, "what will confront him in this mysterious new world? Will any of his conceptions of strange and exotic Martian life prove to be true? Will he find the remains of a long dead civilization? Or will the more conservative opinions of present-day science be borne out with the discovery of a cold and barren planet where only a low form of vegetable life struggles to survive? These questions will be answered by our space pioneers of the future. And they are the ones posed in *Mars and Beyond*. In Disney's finest contemporary style, the sound scientific speculations of the universe are intriguingly probed. Via live-action, animation, and a unique combination of both, the fascinating, exciting story of outer space travel and possibilities of life on other planets constitute an absorbing viewing hour. Introducing the show, Disney says simply, 'One of the most fascinating fields of modern science deals with the possibility of life on other planets. This is our story.' " Technical consultants to the Disney production team in this hour-long, $400 thousand effort were Dr. Earl C. Slipher of the Lowell Observatory, and a former assistant to Percival Lowell; Dr. Ernst Stuhlinger, an atomic physicist; and, of course, Dr. von Braun. For *Mars and Beyond*, however, von Braun collaborated with his old friend Stuhlinger and refined his *Mars Project* armada to a single, atom-powered spacecraft. Assembled in a 1,000 mile-high earth orbit, the ship is 500 feet across and carries a single, reusable landing craft to Mars. The atomic reactor was seen as being housed in the bottom of the gyroscope-shaped ship, and it was

20   Von Braun suggested that the Mars ships would communicate with earth through satellites in earth-orbit. This, of course, was a bold suggestion on the scientist's part, coming, as it did, a half decade before the launch of Sputnik.

200

Above: A superb rendering of a rocket bearing people to Mars in the 1980s. Two nuclear rockets carry the central section aloft, a cylinder consisting of crew quarters and laboratories, a landing module, and a third propulsion system for the return trip. *Below:* When people go to Mars, this Apollo-like system may be the best way of exploring the planet. The interplanetary shuttles with the solar sails will each be capable of bearing 25 tons of material through space. The three-year flight is planned for the 1990s.

MSFC-70-PD-400010

Above: In 1970, this was NASA's plan for a temporary Mars base. Included in the scheme were living quarters for perhaps ten people, a selection of rovers, and a nuclear-powered parent ship in orbit around the planet. *Below:* A semi-permanent base on Mars would probably involve a plant for the production of propellant with which to shuttle between the planet and the parent ship, as well as a means of extracting oxygen from minerals and the atmosphere to keep the explorers alive.

FUEL PLANT

DRILL STATION

MOBILE STATION

CENTRAL POWER SUPPLY

LOGISTICS VEHICLE

MSFC-70-PD-4098

MSFC·70·PD·4097

Above: A permanent Mars base would include this central facility to serve as a link between various outposts. It would also house a medical facility, living quarters, laboratories, and so forth. The thinking at NASA has been to adopt Skylab to serve as just such a Mars base. *Below:* While either a temporary or sophisticated permanent base can and should be established on Mars, vital projects both independent of and related to surface activities will be undertaken in Mars orbit.

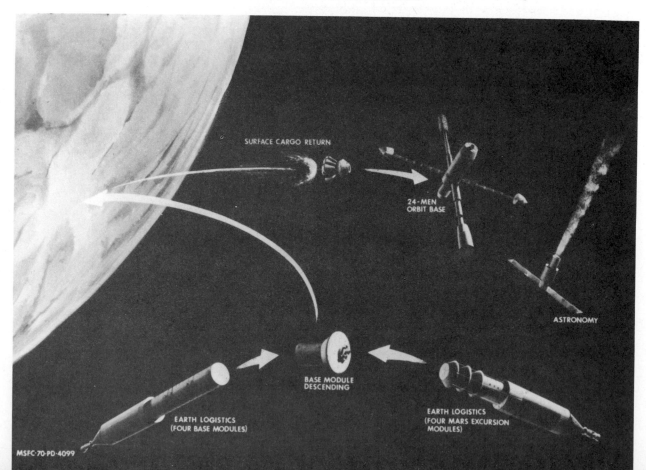

SURFACE CARGO RETURN

24-MEN ORBIT BASE

ASTRONOMY

BASE MODULE DESCENDING

EARTH LOGISTICS (FOUR BASE MODULES)

EARTH LOGISTICS (FOUR MARS EXCURSION MODULES)

MSFC·70·PD·4099

designed to supply a continuous flow of heat which created steam to run generators which, in turn, would power the craft with electricity. The 20-man crew was seen as spending 13 months and six days in transit to Mars, with an additional two months spent in a 620-mile-high Mars orbit before an actual landing.[21]

[However, in addition to its insight regarding a Mars voyage, and the possible nature of indigenous life forms, *Mars and Beyond* made a passing reference to the establishment of permanent colonies on Mars. This is something to which NASA has only recently addressed itself, the agency's proposals being reflected in the illustrations accompanying this text. But with the exception of NASA, very few space observers have ever confronted the psychological as well as the physical problems of such a settlement, be it domed and graceful, à la Disney, or purely functional as in the NASA study. In this respect, scientist Robert Richardson took an interesting overview of one such settlement, a city he called Sylvania (after Rhea Sylvia, the mortal mate of the god Mars). In his book *Mars,* published in 1964, Richardson skillfully crystallized what would be the prime concern of any unadjusted Mars colonist: "Every moment, they would be oppressed by their terrible sense of isolation. The thought would keep intruding *we are the only human beings on Mars, the only human beings on an alien, hostile planet where creatures like ourselves were never meant to exist!*" Richardson maintained that this realization would be followed by the classic isolation syndrome: a deterioration of personality, intellectual lassitude, a taste for unchallenging literature, and mounting irritability. However, he offered some practical advice on how to combat these symptoms. He suggested, most importantly, that both men and women be sent to Mars, and that they partake in regularly scheduled social events such as dances. He also recommended colorful clothing for all colonists, to contrast with the drab Martian environment, as well as regular exercise and athletic routines to keep the muscles strong under Mars' lesser gravity. Finally, he advised a regular changing of personnel, to keep earth people on Mars from going "stir crazy."[22] Whether or not NASA takes

21 Today, several scenes from the *Mars and Beyond* show are a part of the *Trip to Mars* attraction at Disneyland.
22 Richardson has a joke about the proverbial green man from Mars. He notes that by the time an astronaut could be trained for and reach Mars, his or her enlistment for duty would be up. As a result, the planet would host largely "green" men from earth.

any of Richardson's sound suggestions remains to be seen. Indeed, even the matter of a manned trip to Mars remains a question, as the expense of such an undertaking would dwarf even the $30 billion moon program. At the moment, NASA has no concrete plans for Mars beyond the Viking probes, and no proposals other than the sketchy ideas outlined in the illustrations shown on these pages. The space agency's current order of business is to perfect the long-in-production space shuttle, a payload-bearing craft that rides a one-shot booster piggyback into space, then glides back to earth for reuse. The first such craft, *the Enterprise* (named in honor of the star ship seen on the *Star Trek* television series), was completed and unveiled in the summer of 1976, and is now being readied for active space duty. It will put habitable laboratories into orbit, haul sections of permanent stations into space, and, perhaps, as predicted by Wernher von Braun, one day bring parts and people together in earth orbit for a manned trip to Mars. Looking ahead, what will such an expedition be like? At this time, we can only guess that both men and women will partake in the multi-year journey. This, of course, will require a reevaluation of our concepts of morality: it is unlikely that everyone onboard will be married to someone else onboard, just as it is unlikely that they will all remain celibate for the duration of the trip. In addition to equal representation of the sexes, it is certain that a Mars crew will represent many of the world's nations.[23] This international aspect of the flight would not be an idle gesture of hands-across-the-sea diplomacy: in order to pay for their voyage, it is almost a foregone conclusion that all of the participating countries will have to foot a part of the bill. But consider the rewards. If, when all our ambassadors, wars, and blustering politicians have failed, the space program can show humankind how to cooperate in the quest of a common goal, then it will have well-served our race no matter what the cost.

23 One young space buff wrote to NASA in the late 1960s and made the following suggestion for a Mars probe: "I would like to volunteer to ride on your first spaceship to Mars. I weigh only 60 pounds and am an observant boy. I would not marry any of the women up there because I am not fond of girls, any kind or shape. P.S.: I think you are depending on men with bald heads who are old like my dad. I have lots of hair and am young. My brain will stay warm and function better up there where it is cold." Another would-be astronaut wrote, "My friend and I think you should send kids to Mars, because if there is something dangerous there, you wouldn't have to waste spacemen. Besides, I always wanted to be great." It's difficult to argue with logic like that!

7

LUST FOR LIFE

It seems almost a paradox in these days of manned earth-orbit stations and the exploration of space by mechanical proxy, that mythology still has a home in the heavens. Of course, no longer is it the fantasy of Zeus or Hercules, but these contemporary beliefs are no less important to their many devotees than the great gods and heroes were to the ancients. What are these sky cults? Broadly, they are the lore of astrology, flying saucers, and aliens as deities, manifestations of our desperate wish for there to be something more to the universe than just the data found by scientists; impassioned sects which share the primary goal of trying to make a healthier accounting for human-kind than what cold logic tells us is a very minor bustling in a remote corner of a small galaxy.[1] However, as inevitable as this search is the fact that Mars plays a prominent role in its many phases.

[Astrology is a pet peeve of scientists because it is an irrational process that has traditionally been passed off as calculable. As science and science fiction writer Isaac Asimov points out, "It passes human understanding to suppose that the vast universe we now recognize is arranged only as a key for our own insignificant dust-speck." Of course, it has been argued that while the stars and planets may not have been arranged for our benefit, they influence us just the same; perhaps this is so, to a minute degree, the way the moon's

1 Religion is not on the list because it is essentially a nonplanetary form of spiritualism, and thus not directly relevant to our study.

207

gravity unsettles those who are susceptible to such an influence. However, astrology does not even reflect the totality of the universe: where are the black holes, quasars, pulsars, and asteroids in our horoscopes? As Carl Sagan writes in his book, *The Cosmic Connection*, "Astrology has not attempted to keep pace with the times. Even the calculations of planetary motions and positions performed by most astrologers are usually inaccurate."[2] Yet, even if the astrologer's tools were completely up to date, it is the specificity of astrological observations that makes it sheer nonsense and a commerical sham. To imagine that one can assess the wisdom of a financial deal or a decision in love based on our perspective of the planets, stars, and comets is simply absurd. And, while there is no doubt but that many astrologers are sincere, as Isaac Asimov further observes, "I am quite certain that with very little ingenuity, I could invent a chain of reasoning connecting the pattern of burping of a herd of hippopotami amid the reeds of the River Nile with the rise and fall of the steel output in the mills of Gary, Indiana." Nonetheless, many people hold to astrological predictions with honest fervor, and they view the planet Mars as one that greatly affects their lives.

[It is the professed job of the astrologer not to make predictions, but rather, based on the consideration that each person is a microcosm of the universe, to judge the harmony of the forces which appear in his or her natal horoscope. This situation is then interpreted in light of the transiting planets, and a horoscope can be drawn. We mentioned horoscopes in chapter one as they pertained to early astronomy: this is the diagram showing the design of the zodiac as it applies to the individual. The zodiac, you will recall, is the astrological map of the 30-degree area of space through which the planets move, and consists of 12 signs or houses: Aries the ram, Taurus the bull, Gemini the twins, Cancer the crab, Leo the lion, Virgo the virgin, Libra the scales, Scorpio the scorpion, Sagittarius the archer, Capricorn the half-fish/half-goat, Aquarius the water-bearer, and Pisces the fish. In terms of the forces which allegedly affect us all, these constellations provide a general boundary for the astrologer's work; on a particular horoscope, the planets describe the specifics. And as far as astrologers are concerned, Mars represents aggression, energy,

2 In *Write Your Own Horoscope*, renowned astrologer Joseph F. Goodavage states, in reference to Phobos and Deimos, "Ancient astrologers knew of their existence, sizes, speeds, and altitudes from the Red Planet." In fact, they did not.

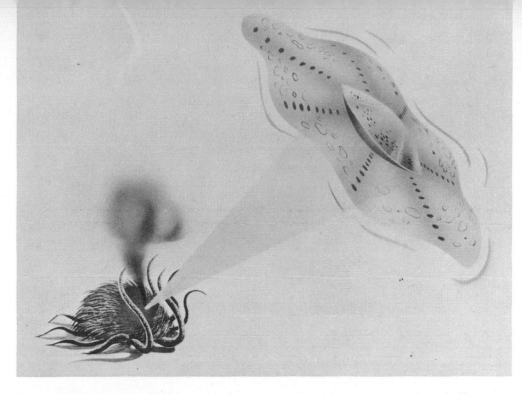

Pages 209–19: A variety of Martian life forms, some fanciful, some practical, and all featured on *Mars and Beyond. (Copyright by Walt Disney Productions).*

power, and action, and is symbolized by the shield and the spear, the same mark used by psychologists to signify the human male. Thus, inconsiderate of your own sign, in whichever house Mars was present at the time of your birth, the planet affects you as follows:[3]

[MARS IN ARIES: If you have Mars in "The Sign of the Self," its presence will be felt by causing you to be enterprising and egocentric, with the need to be without any physical or mental restrictions whatsoever. However, there is every indication that such people are accident prone and may do particular damage to their head and face. They are frank, hot tempered, adventurous, sportsminded, and vitally independent.

[MARS IN TAURUS: As a result of determination and organization, Mars means money and material wealth if you're born when it's in this position. Unfortunately, marriage is wont to be an unrewarding experience.

[MARS IN GEMINI: A sharp, restless mind and mechanical skills (suggesting practicality) are the legacy of having the Red Planet amidst the twins, lending weight to the occupations of lawyer, teacher, or writer. However, as one astrologer notes, people with Mars

3 If you'd like to know your Mars sign, please write to us informing us of your exact birthdate and enclosing a stamped, self-addressed envelope.

in their Gemini should beware: "Nervousness may cause you to smoke excessively, which will lead to pain in your lungs."

[MARS IN CANCER: Bad eyesight is in the stars for this juxtaposition, as well as ulcers, problems with your mother, the prospects of a bad marriage, and the chance that your residence will be smitten with natural disaster. However, don't pack it all in: if you survive these problems, things look pretty good due to your toughness and determination.

[MARS IN LEO: Like those with Mars in Cancer, you, too, have your problems! Because of your superior muscular strength, you are liable to overextend and thus cut yourself, run a fever, or entertain a mishap with fire. However, this is the price you pay, since you love winning, power, and all other ego-boosting stimuli. You also like to gamble in business and have variety in love, usually succeeding at both; you have, respectively, a good mind for investment or gaining the trust of those in authority, and are kindly and affectionate.

[MARS IN VIRGO: If ever there were two incompatible signs—the soldier and the virgin—these are they! Although you want some measure of power and notoriety, failure will litter your course, and you will become physically ill from overwork. You are liable to argue with and alienate friends, or become obstinate and irritable at all the wrong times. However, the astrologers tell us that, in the end, you are careful about your health and diet . . . even though you are likely to suffer from gastric problems in any case.

[MARS IN LIBRA: First and foremost, you like variety in your love life, and must be in the midst of a social whirl. There will be tough competition in business, and the indicators are not good for the acquisition of a top-drawer education until late in your life. But don't worry: when you eventually marry, the signs are good that at least your children will be smart. .

[MARS IN SCORPIO: You are made of intensity, grit, and determination, and spend much more time diligently at work than relaxing. Mars in the scorpion also gives you great creative sting, thunderously active emotions, and an interest in mysticism and astrology. You have the potential to hold a grudge and be a cynic—although this aspect of your personality is easily offset by the presence of other planets in your horoscope. A leading astrologer lists surgeon, oil driller, deep sea diver, or prize fighter as your most promising occupations.

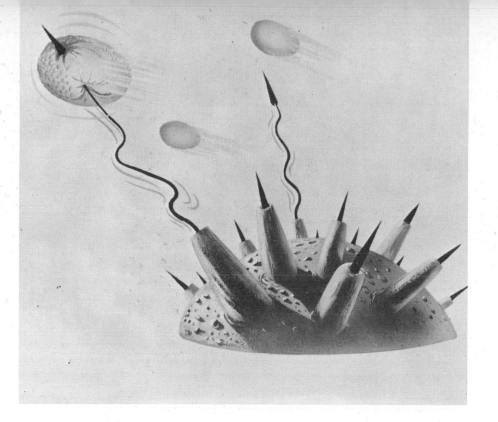

[MARS IN SAGITTARIUS: Shed your current guise: you are, in effect, the Lone Ranger! You are righteous and brave, love a long journey and high adventure, and bow to no one. You are generous, love the outdoors, and find inspiration in military maneuvers. You may not stumble upon a silver mine like the illustrious Ranger Reid, but you will certainly earn your money through inheritance or marriage.

[MARS IN CAPRICORN: To put in bluntly, you're a martyr. You bear heavy burdens, including those which are not your own, and do so with a smile. You are accident-prone, with your legs a particularly weak area, but there will be those who are only too willing to help you back to your feet, since you are admired for your altruism and ready acceptance of both hard work and responsibility.

[MARS IN AQUARIUS: Although the union of the war and water forces would seem to predestine you for service in the Navy, such is not the case. Your ideal occupation is as a politician, a social worker, or a physician. Not only do you have a logical, lightning-fast mind, but you are a crusader for the rights of others—particularly minorities— and will achieve considerable success in this area.

[MARS IN PISCES: This one is another hard-luck conjugation, as you are irresolute, lack character, and slip easily into melancholia. You

211

will be frustrated in love, often strive to help others at great cost to your own well-being, and should not handle liquids—with which you are unfortunately very clumsy. You are also a daydreamer, and find it difficult to keep your mind on one thought at a time. But don't give up hope! You are fated to tie in with some wealthy and influential people, who will be only too happy to help you both professionally and financially.

[For your further fun and elucidation, here is how some well-known personalities fit into this scheme of things: John F. Kennedy, Mars in Aries; football player Joe Namath, Mars in Aries; Richard Nixon, Mars in Sagittarius; actor Charlton Heston, Mars in Virgo; Ex-Beatle Paul McCartney, Mars in Leo; author Norman Mailer, Mars in Aries; Henry Kissinger, Mars in Gemini; and Walt Disney, Mars in Capricorn. While several of these associations are impressively accurate—particularly where Kennedy is concerned—each of the men can also be categorized in other Mars signs, with equally appropriate results. To wit: try Nixon as Mars in Pisces and Disney as Mars in Gemini.

[When we one day colonize Mars, it will be interesting to see how the astrologists classify earth in these constellations—especially when the signs will no longer look exactly the same as they do from our native world. Of course, with two moons in the sky rather than one, astrological computations may become too complex to be worth anyone's while. For the present time, then, this pseudoscience gives solace to those who simply *must* have a transcendental link with the stars. Eventually, however, when knowledge will have replaced the various superstitions and fears which keep astrology alive, it will be remembered as a curious, long-lived terrestrial fad . . . and nothing more.

[Speaking of fads, Ken Arnold initiated one when on June 24, 1947, while piloting his private plane over the state of Washington, he saw nine round objects—"flat, like pie tins," as he described them—and thus began the flying saucer craze. Within weeks, and through the years to follow, thousands of people saw alien vessels, and those who weren't actually sighting them were busy *riding* spaceships to various interplanetary points. Among the most famous such passengers were George Adamski and Cedric Allingham, both of whom had experiences with Martians. Adamski, the more widely traveled of the two, wrote about his voyages in the book *Flying Saucers Have Landed* (1953) and *Inside the Space Ships* (1955), although his contact with residents of the fourth planet consisted of nothing more than a tour of one ship conducted by a Martian member of the multiplanetary crew. The alien explained the workings of the lampshadelike spaceship to Adamski, describing the three bubbles underneath the craft as the vehicle's power sources, spheres which collected and condensed static electricity. Allingham's encounter with a Martian was more involved, outlined in his book *Flying Saucer From Mars*. On a bird-watching vacation in Scotland, Allingham saw the titular disc land and was greeted by an alien wearing a one-piece space suit. The visitor had two tubes inserted in his nostrils to help him breathe the thicker air on earth; he spoke no English, but managed to convince Allingham to go for a trip in his saucer. The Briton obliged, was duly returned, and the entire event was witnessed by lighthouse keeper James Duncan, who signed a statement for Allingham vouching as to what he had seen. The fact that no one could find a trace of this James Duncan leaves the validity of his testimony somewhat in doubt.

[Through the years, dozens of people have claimed to have flown in flying saucers, and their accounts—many of which are intricately spun—are obviously false, created for publicity and sensationalism. However, both Adamski and Allingham actually *believe* that they had

213

been the guests of Martians. But there has never been any hard corroborative evidence to support their stories, or the existence of flying saucers themselves, so the reader is invited to draw his or her own conclusions on the subject. However, this entire question of alien visitation has given birth not only to the flying saucer phenomenon, but to the idea that beings from outer space are somehow watching over us. The Aetherius Society claims to have an inside track on the situation. Its founder, George King, is earth's representative in an organization called the Interplanetary Parliament, which meets periodically on Saturn. Governed by a Venusian known as the Master Aetherius, an example of the body's circum-system duties include a recent instance wherein they thwarted an invasion of earth by a race of fish people living on the far side of the galaxy. The world of these aquatic creatures was fast losing its seas, so they fired a poison bomb our way, intending to kill off humanity and seize our fertile home. Fortunately, the Martians were on their toes (or tentacles, as the case may be) and intercepted the deadly projectile with a thunderbolt. In fact, heading this particular operation was none other than Jesus Christ, who currently resides on Mars. According to the Aetherius Society, the Nazarene was originally a Venusian who came to earth in a flying saucer—mistakenly referred to as the Star of Bethlehem—and, when he ascended, took up with the Martians. What is the Mars of Aetherius? Well, according to Mr. King, the Martians were originally earthlings, inhabitants of Atlantis, who zipped into space when the continent sunk. They founded the industrial planet of the solar system: Mars boasts immense saucer factories, with all the manual labor done by thought-controlled robots. What's more, the Martians are able to teleport themselves through solid matter, or can move objects across the galaxies. They have superb schools built of crystal, which fertilize young Martian brains with magnetic vibrations from all points of the solar system. And, although they live underground, the Martians frequently surface to eat vegetation which grows along the canals (sic).[4] Beyond their peacekeeping and industrial chores as members of the Interplanetary Parliament, the Martians also maintain a number of important space installations. Mars Sector Eight, the

4 According to the Aetherians, the reason that Viking did not find canals is because the Martians did not want it to. They fed it imaging data which they felt was appropriate. In the past, they have been more generous with us, allowing American astronauts to land on the moon. Russian probes have usually been destroyed.

Governor of the artificial Mars satellite Phobos, uses his station to control both the weather on Mars and serve as a relay point for interplanetary communication.[5] Mars Sector Six is the being responsible for two artificial earth satellites. These vessels, which are oval, immaculately clean, and scented with exotic perfumes, use large prismatic crystals to gather select spacial radiation and isolate the so-called "universal life forces," training them on earth for the betterment of humankind.

[Certainly, the Aetherius Society is what we shall generously call one of the more *imaginative* groups involved with extraterrestrial life and its influence on earth. However, there are scholars whose work is more competent and whose theories are substantially more popular than those of Mr. King, even though, in the end, they are no less improbable. One of the most iconoclastic spokesmen for interplanetary phenomena is Dr. Immanuel Velikovsky. While the Soviet scientist

5 The argument for an artificial Phobos was more sensibly presented by Russian astrophysicist I. S. Shklovskii in 1959. He suggested that the moon might be hollow because its average velocity was ever-increasing. He took this to mean that the satellite was falling slowly toward Mars, something which would usually be the result of atmospheric drag. However, because the Martian atmosphere is so tenuous, Shklovskii decided that, instead, Phobos must be extremely light—perhaps a hollow sphere launched by Martians. Further observations proved that the increasing velocity existed only in the scientist's slightly inaccurate computations.

had done his fair share of respectable research during the early years of his career, it was in the field of psychiatry, not astronomy. However, he produced a handful of postulates which were published as the book *Worlds in Collision* (1950) and caught the public's fancy. In this best-selling work, Velikovsky claims that the miracles of the Bible are the result of stellar events. For instance, his most popular claim is that Venus was once a piece of Jupiter which, when torn from its parent planet, left the mysterious Great Red Spot.[6] Passing the earth, Venus caused it to stop dead in its tracks, leaving the Red Sea dry so that Moses could pass through; when Venus moved on, the earth began to rotate again, drowning the soldiers of the Pharoah. The planet's comet-like tail caused manna to rain on the wandering Hebrews; bouncing back and forth in an unsteady orbit, it was even Venus, and not Joshua, that caused the walls of Jericho to come tumbling down. And, before it stopped its intersystem bucking, Venus managed to collide with Mars, causing the Red Planet to move closer to the earth, nearly hitting it in 687 B.C.; Mars' fluctuation is thus described as having been responsible for many of the Old Testament disasters. The problem with Velikovsky's colorful arguments is that they all depend upon physical impossibility. For one thing, from where did all that energy suddenly come to rip out a large piece of Jupiter? The Master Aetherius? For another, why is there no mention, in the records of cultures contemporary with the Exodus, of planet-wide catastrophes which must surely have resulted from the abrupt stopping and starting of earth. And, most significantly, there is no way that a body the size of Venus could possibly have passed by and simply halted the earth. The physics are incongruous. But the hypotheses of Velikovsky still hold a great deal of influence with the lay public, in part because the Russian uses quasi-rational reasoning to explain some of history's greatest puzzles, and also because he has been attacked by scientists for being both an outsider to the field and spinning fairy tales to boot—at a time when people are generally anti-science. Thus, while Velikovsky does not fit implicitly into our outline of thinkers who look to strengthen the spiritual bonds between humankind and the rest of the universe, he is clearly of the ilk to which Isaac Asimov made reference in his hippopotami observation.

[Despite the popularity of Velikovsky, the publishing industry has

6 The Great Red Spot, over seven times the size of Venus, is an unchanging blob of red in the atmosphere of Jupiter. It is thought to be a perpetual storm system.

not seen an explosion of extraterrestrial nonsense to match the wake caused by Erich von Daniken's *Chariots of the Gods*. First there were the author's own sequels *Gods from Outer Space*, *Gold of the Gods*, *Miracles of the Gods*, etc., and then the many non-von Daniken tomes which were quickly written to hop on the bandwagon, books such as *Gods, Demons and Space Chariots*; *God Drives a Flying Saucer*; *The Coming of the Gods*; *The Home of the Gods*; the anti-von Daniken *Crash Go the Chariots*; and so forth. The basic premise of these many volumes is that, at some time in the past, alien beings visited the earth and either gave birth to or became the guardians of the human race. Some of the authors claim that the Gods of Olympus were aliens, conventioneering or stranded on earth; others believe that Moses, Jesus, Leonardo da Vinci, and Jules Verne were all the same man, a being who assumed various disguises through the centuries to help our world. And the concensus among these scribes is that if Mars were not the prime source for these visitations, it is certainly in the running! Von Daniken is particularly fond of the argument that Martians came to earth in the remote past and bred with the semi-intelligent creatures they found living here. Due to the lower gravity of Mars, these creatures, of course, were giants, and were able to move the great rocks found at such sites as the Pyramids of Egypt;

217

Sacsahuaman, Peru, where walls and what von Daniken thinks is a giant throne, were built from huge blocks; Easter Island, where there are large stone sculptures; etc., the transportation of which von Daniken claims was beyond any terrestrial means of the times. The Martians also instructed us in art and other than motor sciences, eventually dying out and leaving behind enlightened Homo sapiens where once ruled the apes. Von Daniken supports his theories based largely on the belief that ancient art shows space-suited beings, and that marks on our plains and in mountainsides were landing strips or aerial markers used by extraterrestrial astronauts. Once again, the facts are what the author choses to make of them. Von Daniken misquotes his source material (according to Carl Sagan, who was himself misrepresented) and, as Ben Bova so neatly summed it up in an editorial in the June 1975 issue of *Analog Science Fiction/Science Fact* Magazine, "The evidence seems very substantial, and all stacked against von Daniken. The miracles of engineering and monument building that he ascribes to the Chariot-riding visitors were quite clearly built by humans, for humans. There is written evidence showing how the pyramids were built; there are contemporary demonstrations available for anyone who seriously wants to know how the Easter Island statues were set up, how the Stonehenge monoliths were erected, and how our Cro-Magnon forebears decorated their caves. There is no discernible need for antigravity stone carriers, or laser-cutting tools." All of which tells us, in no uncertain terms, that outer space is the rightful realm of humanity, in terms of the future, and of science for the present. Yes, astronomers have theories; God knows, Percival Lowell had them! But *questioning* is the nuts and bolts of any scientific endeavor. And, since the purpose of science is, by and large, to accumulate and interpret facts, not dollars, hypotheses which are proven by subsequent findings to be inaccurate are *dropped*. Cold. This is what distinguishes the astrologers, the Aetherians, and the Velikovskys from the people at NASA and the Viking staff. There will always be room for science fiction as well as liberal scientific speculation; it flexes the imagination and keeps it supple. But if such writing is liberal with the *facts*, it will, in the long run, do more harm than good for the exploration of space. For instance, because the von Daniken school has told us that there are alien beings literally knocking on earth's front door, it is the scientists, and not the irresponsible fantasists, who look foolish when Viking has

trouble—and rightfully so; it's a tough search by remote control—distinguishing Martian chemistry from biology. Thus, while the scientists may not make the front page as often as the sensationalists, they are certainly building more soundly toward the future. Which brings us to the inevitable question: are there extraterrestrials? The mathematics certainly favor life as a recurring rather than an isolated phenomenon. The fact that we may not find any organisms on eight of our solar system's nine planets in no way indicates that life does not exist in the solar systems which undoubtedly circle the trillions of other stars in our universe. Even if the ratio were, as it is in the sun's family, at least one in nine, that would still give us hundreds of billions of inhabited planets. Unfortunately, if Mars and Titan prove to be dead worlds, then the chance of uncovering alien beings in our own lifetimes is remote; the distances between the stars are simply too great.[7] Let us suppose, however, that we *do* find life in space sometime during the next few decades, be they primitive microbes or creatures far in advance of human beings—of the Valentine Michael Smith sort. What will it mean to our existence? To put this hypothetical situation in a cultural perspective, I canvassed religious leaders for their opinions. For, in the event of communication with beings from outer space, the various faiths in our Judeo-Christian society would be called upon by the non-science oriented public for answers—and, perhaps, reassurance.[8] "There might be fear," I was told by a priest in Manhattan, "and skepticism, I'm sure, if these beings were more intelligent than we. But if they would welcome us, then everything would be okay. People have become accustomed to accepting new things in our world." A Los Angeles priest concurred, but admitted that, "although there *has* been a lot written about the possibility of life on Mars and elsewhere, I'm afraid that people don't generally read, and I would not be surprised if there were fear at first." Both men said that they don't believe the earth is the only planet that was created for habitation. Regarding the advantages of such interstellar communication, the clearest idea of its potential to benefit humankind came from Philadelphia's Rabbi Yaakov Rosenberg: "Personally, I look forward to contact with other life forms. It would show us that this is not the only way to live. We've become too

7 With the exception of the sun, the star nearest to the earth is Alpha Centauri, which is 4.3 light years away.
8 With the exception of Rabbi Yaakov Rosenberg, all of the men who responded to my questions wished to remain anonymous.

The breathtaking sunset at Chryse Planitia. Barely visible in the foreground left is one of Viking's power system covers. The panorama covers 120 degrees.

cynical in our world, and we need something like this to excite us. And we can conceivably learn from these people. We can try not to make mistakes that they made. After all, before the earth came into being, we are told that God created and destroyed many worlds." All of the clergymen with whom I spoke agreed that there would be no need to reassess the validity of the Bible or of the Biblical idea that humankind was created in God's image, since they all interpret "image" as a figurative reference to spiritual rather than physical being. "All that matters," Rabbi Rosenberg concluded, "is that they believe in the notions of right and wrong."

[Perhaps it's only fair that we give a closing comment on the matter of alien life forms to the articulate Percival Lowell, the man who gave both the exploration of Mars and the idea of extraterrestrial life its greatest impetus. In *Mars*, he wrote:

> That Mars seems to be inhabited is not the last, but the first word on the subject. More important than the mere fact of the existence of living beings there, is the question of what they may be like. Whether we ourselves shall live to learn this cannot, of course, be foretold. One thing, however, we can do, and that speedily: look at things from a standpoint raised above our local point of view; free our minds at least from the shackles that of necessity tether our bodies; recognize the possibility of others in the same light that we do the certainty of ourselves. That we are the sum and substance of the capabilities of the cosmos is something so preposterous as to be exquisitely comic.

[In short, the future bodes exciting for humankind on Mars and elsewhere in space. There is no need for metaphysics; science is profound enough as is! The trick is to make it not just the matter of a few dollars grudgingly given scientists with which to build rockets and robots, but an adventure for all people. There is nothing to lose—and the universe to gain!

APPENDICES

APPENDIX I

[At 12:00 noon on September 4, 1976, the following conversation took place between then–President Gerald Ford and Viking executives Jim Martin and John Naugle. They were discussing the landing of Viking II.

NAUGLE: Yes, Mr. President.

FORD: How are you?

NAUGLE: We're fine. Very happy out here.

FORD: I should think you should be, and let me congratulate you and the NASA Viking team of Universities and Industry and the NASA organization for a first-class job.

NAUGLE: Thank you, Mr. President. There are about ten thousand people that are very grateful for those words, and very grateful for the support of the American people that made this mission possible.

FORD: Could you tell me a little bit about it? I know you had some radio communication trouble at the outset, but I understand that that has been remedied and you do have two pretty good pictures.

MARTIN: Yes, Mr. President. We had some initial difficulties, yesterday afternoon after separation of the lander from the orbiter. We lost communications between the orbiter and earth for a while. When we were able to—about an hour later—we were able to reestablish those communications, but we were in a low gain antenna mode at a lower signal strength, and as a result, we were unable to monitor the progress of the lander as it descended to the Martian surface. We, in effect—the lander, which is a very superb automated machine, landed and did everything on its own, and landed right on schedule. But we were sitting somewhat on pins and needles because of the fact that we had no information here on the ground.

FORD: Well, I think that was a tremendous feat, that something that far away would do the job all by itself without human hands, at least at the present time, having anything to do with it.

MARTIN: As Dr. Naugle said, sir, I believe the ten thousand human hands that have worked seven years are what made it all possible.

FORD: Well, I think this is a superlative demonstration of American skill and teamwork, and I think the world will recognize it and I certainly want to congratulate all of the team for a great, great job. Can you give me a little information about this landing site, vis-a-vis, one where the first landed?

MARTIN: Yes sir. We are sitting on a plain that is called Utopia.

FORD: (laughter)

MARTIN: Let me say that I have been promised, when I selected this landing site, that we would see a lot of sand dunes. And perhaps we're *seeing* a lot of sand dunes . . . but they're now covered with rocks.

FORD: I see.

MARTIN: We have very superb pictures—beautiful pictures. Very sharp—very clear. But there are an awful lot of rocks. We're about four thousand miles from the landing one site. We're at forty-eight degrees north latitude, whereas the earlier site is closer to the equator.

FORD: I see.

MARTIN: We are hopeful that this site is in an area where we have measured more water in the atmosphere and therefore we are hopeful that we'll find perhaps more water in or near the surface and this might be more conducive to life.

FORD: I saw on the eleven o'clock news in Washington, last night, all of you congratulating each other, and I was there in spirit if not in person.

MARTIN: Well, thank you sir. We really had a tremendous team, and not only the Americans, but I'd like to recognize the fact that we have colleagues in Australia and in Spain that operate tracking stations of the Deep Space Network that are a vital part of this operation.

FORD: Well, give them my very best as well. I know you're in communication with them. Let me ask a question for the future. What other planetary exploration do you have in mind?

NAUGLE: Mr. President, we have a mission scheduled for—to go to explore Saturn, starting next spring, and then later, a—in 1978, a mission to Venus to make a more detailed exploration of the atmosphere of Venus. We do not have anything in the pipeline beyond those two, and we are in the process, as I'm sure you're aware, of defining what the future United States planetary exploration program should be.

FORD: Right.

NAUGLE: In the case of Mars follow-on, we had waited on that until we had a—some understanding of what we learned from Viking. As you might expect, Mr. President, there's a great deal of discussion out here about what kind of a Mars follow-on program there should be. There are people who are saying that we ought to take this magnificent Viking machine and put it on some kind of rollers and move it so we can move up on the hill and get a better view of Mars, and explore some of the great canyons that we see there. There are other people who say that the surface chemistry, as I'm sure you've seen in the papers—we're very puzzled by the measurements that the biology experiments and the other experiments are making of the nature of the surface of Mars. And there are people who say that we should come up with new, better, more sensitive experiments to study that. And there are other people who say we should bring a sample of Mars back to laboratories here on the earth so that we can make a more detailed, thorough analysis. There's quite a sense of excitement out here. To me, it's comparable to that back in the Sputnik era. Except that I think there's more solid intellectual curiosity on the part of the public—more of a feeling of really wanting to understand what the nature of this planet, Mars, is. More so of that than a worry about our competitive position with respect to the U.S.S.R.

FORD: Well, it is very, very exciting, and brings back memories when I was on an aircraft carrier in World War II, when I was the Assistant Navigator and we used to take our readings at sunrise and sunset—and they never—they always said Mars was totally unreliable. Is that true in navigation?

MARTIN: No, sir. We found it exactly where we predicted it would be. You know, these two spacecraft traveled about 440 million miles—between earth and Mars—and they arrived right on time and right where we expected. We have a superb navigation team, and I think that's one thing we can do pretty well is get on the planet. I must admit, we are all somewhat amazed everytime we see a new picture, because we don't quite understand what's going on here on the surface.

FORD: Well, it's a great accomplishment, and, again, congratulations, and I'll be hearing from Jim Fletcher, I'm sure, and we look forward to working with you, and you've all done a great job, and all of us are very proud of you.

NAUGLE: Thank you, Mr. President.

MARTIN: Thank you, Mr. President.

FORD: Have a good weekend.

NAUGLE: We will.

MARTIN: Thank you.

APPENDIX II

[The following firms were Viking contractors, responsible for building the various space craft components.

ORBITER
Prime Contractor:
Jet Propulsion Laboratory
Subcontractors:
Martin Marietta Aerospace (Propulsion Systems)
Rocketdyne Corporation (Propulsion Engines)
General Electric Co. (Attitude Control System)
Honeywell Radiation Corporation (Celestial Sensor)
Motorola Inc. (Relay Radio and Telemetry; Radio Subsystem)
Philco-Ford Corporation (S-Band and Relay Antennae)
General Electric Co. (Computer Command System)
Motorola Inc. (Flight Data Subsystems)
Texas Instruments (Data Storage Subsystem; Electronics)
Lockheed Electronics (Data Storage Subsystem; Transporter)
Electro-Optical Systems (Xerox Corporation) (Power Subsystem)

SCIENCE INSTRUMENTS
Subcontractors:
Ball Brothers Research Corporation (Orbiter Imaging; Visual Imaging
    Subsystem)
A.T.C. (Water Vapor Mapping; Mars Atmosphere Water Detector)

Santa Barbara Research Center (Thermal Mapping; Infrared Thermal
Mapping)
Bendix Aerospace Systems Division (Entry Science; Upper Atmosphere Mass
Spectrometer; Retarding Potential Analyzer)
Hamilton Standard Division, United Aircraft (Lander Accelerometers)
K-West Industries (Aeroshell Stagnation Pressure Instrument)
Martin Marietta Aerospace (Recovery Temperature Instrument)

LANDER
Prime Contractor:
Martin Marietta Aerospace
Subcontractors:
Schjeldahl Inc. (Bioshield)
Martin Marietta Aerospace (Aeroshell)
Goodyear Aerospace Corporation (Parachute System)
Rocket Research Corporation (Landing Engines)
Celesco Industries (Surface Sampler)
RCA Astro-Electronics Division (Communications)
Honeywell Inc., Aerospace Division (Guidance, Control and Sequencing
Computer; Data Storage Memory)
Martin Marietta Aerospace (Data Acquisition and Processing Unit; Landing
Legs and Footpads)
Teledyne Ryan Aeronautical (Radar Altimeter and Terminal Descent and
Landing Radar)
Energy Research and Development Administration (Radioisotope
Thermoelectric Generator)
Lockheed Electronics Co. Inc. (Tape Recorder)
Hamilton Standard Division, United Aircraft (Intertial Reference Unit)
General Electric Battery Division (Batteries)

SCIENCE INSTRUMENTS
Subcontractors:
Itek Corporation (Lander Imaging; Facsimile Camera System)
TRW Systems Group (Biology Instrument)
Litton Industries (Molecular Analysis; Gas Chromatograph Mass
Spectrometer)
Martin Marietta Aerospace (Inorganic Chemistry; X-Ray Fluorescence
Spectrometer)
TRW Systems Group (Meteorology Instrument System)
Bendix Aerospace Systems Division (Seismometer)
Celesco Industries (Physical Properties; Various Instruments, Indicators,
Mirrors)
Raytheon Inc. (Magnetic Properties; Magnet Arrays)

LAUNCH VEHICLE
Contractors:
Martin Marietta Aerospace (Titan III-E)
General Dynamics/Convair (Centaur)

APPENDIX III

[The following people comprised the various Viking Science Teams.

Orbiter Imaging:
Michael H. Carr (U.S. Geological Survey)
William A. Baum (Lowell Observatory)
Karl R. Blasius (Science Applications Inc.)
Geoffrey Briggs (JPL)
James A. Cutts (Science Applications Inc.)
Thomas C. Duxbury (JPL)
Ronald Greeley (University of Santa Clara)
John E. Guest (University of London)
Harold Masursky (U.S. Geological Survey)
Bradford A. Smith (University of Arizona)
Lawrence A. Soderblom (U.S. Geological Survey)
Joseph Veverka (Cornell University)
John B. Wellman (JPL)
Thermal Mapping:
Hugh H. Kieffer (University of California, Los Angeles)
Stillman C. Chase (Santa Barbara Research Center)
Ellis D. Miner (JPL)
Guido Munch (California Institute of Technology)
Gerry Neugebauer (California Institute of Technology)
Frank Palluconi (JPL)
Water Vapor Mapping:
C. Barney Farmer (JPL)

Donald W. Davies (JPL)
Daniel D. LaPorte (Santa Barbara Research Center)
Entry Science:
Alfred O. C. Nier (University of Minnesota)
William B. Hanson (University of Texas)
Michael B. McElroy (Harvard University)
Alfred Seiff (Ames Research Center)
Nelson W. Spencer (Goddard Space Flight Center)
Lander Imaging:
Thomas A. Mutch (Brown University)
Alan B. Binder (Science Applications Inc.)
Friedrich O. Huck (Langley Research Center)
Elliott C. Levinthal (Stanford University)
Sidney Liebes, Jr. (Stanford University)
Elliott C. Morris (U.S. Geological Survey)
James A. Pollack (Ames Research Center)
Carl Sagan (Cornell University)
Biology:
Harold P. Klein (Ames Research Center)
Norman H. Horowitz (California Institute of Technology)
Joshua Lederberg (Stanford University)
Gilbert V. Levin (Biospherics Inc.)
Vance I. Oyama (Ames Research Center)
Alexander Rich (Massachusetts Institute of Technology)
Molecular Analysis:
Klaus Biemann (Massachusetts Institute of Technology
Duwayne M. Anderson (National Science Foundation, Office of Polar
    Programs)
Alfred O. C. Nier (University of Minnesota)
Leslie E. Orgel (Salk Institute for Biological Studies)
John Oro (University of Houston)
Tobias Owen (State University of New York)
Priestley Toulmin III (U.S. Geological Survey)
Harold C. Urey (University of California, San Diego)
Inorganic Chemistry:
Priestley Toulmin III (U.S. Geological Survey)
Alex K. Baird (Pomona College)
Benton C. Clark (Martin Marietta Aerospace)
Klaus Keil (University of New Mexico)
Harry J. Rose (U.S. Geological Survey)
Meteorology:
Seymour L. Hess (Florida State University)
Robert M. Henry (Langley Research Center)
Conway B. Leovy (University of Washington)
Jack Ryan (McDonnell Douglas Corporation)
James E. Tillman (University of Washington)

Seismology:
Don L. Anderson (California Institute of Technology)
Fred Duennebier (University of Texas)
Robert L. Kovach (Stanford University)
Gary V. Latham (University of Texas)
George Sutton (University of Hawaii)
Nafi M. Toksoz (Massachusetts Institute of Technology)
Physical Properties:
Richard W. Shorthill (University of Utah)
Robert E. Hutton (TRW Systems Group)
Henry J. Moore II (U.S. Geological Survey)
Ronald F. Scott (California Institute of Technology)
Magnetic Properties:
Robert B. Hargraves (Princeton University)
Radio Science:
William H. Michael (Langley Research Center)
Joseph P. Brenkle (JPL)
Dan L. Cain (JPL)
John G. Davies (University of Manchester)
Gunnar Fjeldbo (JPL)
Mario D. Grossi (Raytheon Corporation)
Irwin I. Shapiro (Massachusetts Institute of Technology)
Charles T. Steizried (JPL)
Robert H. Tolson (Langley Research Center)
G. Leonard Tyler (Stanford University)
Viking Flight Team:
J. S. Martin, Jr. (Project Manager)
A. T. Young (Mission Director)
G. A. Soffen (Chief Project Scientist)
J. D. Goodlette (Chief Engineer)
B. G. Lee (Science Analysis and Mission Planning Director)
P. T. Lyman (Spacecraft Performance and Flight Path Analysis Director)
M. J. Alazard (Mission Control Director)
G. N. Gianopulos (Mission Control Computing Center Systems Engineer)
D. J. Mudgway (Deep Space Network Manager)
H. W. Norris (Senior Staff, Orbiter Operations)
W. O. Lowrie (Senior Staff, Lander Operations)
H. E. Van Ness (Senior Staff, External Affairs)
R. L. Crabtree (Deputy Mission Director)
L. Kingsland (Deputy Mission Director, Planning)
C. W. Snyder (Orbiter Science Group Chief)
G. C. Broome (Lander Science Group Chief)
J. D. Porter (Mission Planning Group Chief)
R. A. Ploszaj (Orbiter Performance Analysis)
R. W. Sjostrom (Lander Performance Analysis)
W. J. O'Neil (Flight Path Analysis)

M. M. Grogan (Sequence Development
L. S. Canin (Flight Control)
D. D. Gordon (Data Support)
W. B. Green (Image Processing Staff Leader)
H. Masursky (Landing Site Staff Leader)
K. S. Watkins (Administrative Support Office)
K. W. Graham (Ground Data Systems Support)
R. J. Polutchko (Lander Support Office Chief)
K. H. Farley (Lander Support Engineering)
F. D. Nold (Lander Support Operations)
B. A. Claussen (Lander Support Software)

BIBLIOGRAPHY

[This bibliography is in three parts. The nonfiction bibliography is a listing of those books which served as the author's primary reference sources. The nonfiction magazine bibliography is recommended for readers who wish to study aspects of the Red Planet in great, often scholarly, detail. The fiction bibliography lists those books which were discussed in this text and are, or have recently been, in print.

NONFICTION

Antoniadi, E. 1930. *The planet Mars*, Paris: Hermann.

Asimov, Isaac. 1975. *Eyes on the universe*. New York City: Houghton Mifflin Co.

Bova, Ben, and Bell, Trudy, eds. 1977. *Other worlds*. New York City: St. Martin's Press.

Branley, Franklin M. 1975. *Mars: Planet No. 4*. New York City: T. Y. Crowell.

Cade, C. M. 1966. *Other worlds than ours*. New York City: Taplinger.

Campbell, Joseph. 1973. *Myths to live by*. New York City: Bantam Books.

Clarke, Arthur C. 1964. *The exploration of space*. New York City: Fawcett.

Cross, Charles A., and Moore, Patrick. 1973. *Mars*. New York City: Crown Publishers.

Glasstone, Samuel. 1968. *The book of Mars*. Washington, D.C.: NASA (SP–179).

Goodavage, Joseph F. 1975. *Write your own horoscope*. New York City: New American Library.

Lewis, Richard S. 1969. *Appointment on the moon.* New York City: Ballantine Books.

Lowell, Percival. 1895. *Mars.* Boston: Houghton Mifflin Co.

Mandel, Siegfried. 1969. *Dictionary of science.* New York City: Dell Publishing Co.

Motz, Lloyd. 1976. *The universe: its beginning and its end.* New York City: Charles Scribner's Sons.

Muirden, James. 1968. *Amateur astronomer's handbook.* New York City: Crowell.

Nourse, Alan E. 1970. *Nine planets.* New York City: Harper & Row.

Richardson, Robert. 1964. *Mars.* New York City: Harcourt Brace Jovanovich.

Sagan, Carl 1975. *The cosmic connection.* New York City: Dell Publishing Co.

Sagan, Carl 1975. *Other worlds.* New York City: Bantam Books.

Velikovsky, Immanuel 1977. *Worlds in collision.* New York City: Pocket Books.

Von Braun, Wernher 1953. *The Mars project.* Urbana, Illinois: University of Illinois Press.

Von Daniken, Erich 1971. *Chariots of the gods.* New York City: Bantam Books.

Wilford, John Noble 1969. *We reach the moon.* New York City: Bantam Books.

MAGAZINES

Analog Science Fiction—Science Fact:

Hoagland, R. December 1974. Why we won't find life on Mars. 51–70

Astronautics and Aeronautices:

Sagan, Carl September 1972. Mars, the view from Mariner IX. 26–41.

Young, Richard S. October 1965. Automated life detection. 70–76.

Icarus (the International Journal of Solar System Studies):

Balsama, S. R., and Salisbury, J. W. Slope angle and frost formation on Mars. 18: 156–163.

Barker, E. S. Martian atmospheric water and vapor observations: 1972–74. 28: 247–268.

Belcher, D., Veverka, J., and Sagan, Carl. Mariner photography of Mars and aerial photography of Earth: some analogies. 15: 241–252.

Carr, M. H., The role of lava erosion in the formation of lunar rilles and Martian channels. 22: 1–23.

Farmer, C. B., Liquid water on Mars. 28: 279–289.

Horowitz, N., Hubbard, J. S., and Hobby, G. L. The carbon-assimilation experiment: the Viking Mars lander. 16: 147–152.

Houck, J. R., Pollack, J. B., Sagan, C., Schaack, D., and Decker, J. A. High altitude infrared spectroscopic evidence for bound water on Mars. 18: 470–480.

Klein, H. P., Lederberg, J., and Rich, A. Biological experiments: the Viking Mars lander. 16: 139–146.

Levin, G. V. Detection of metabolically produced labelled gas: the Viking Mars lander. 16: 153–166.

Lipa, B., and Tyler, G. L. Surface slope probabilities from the spectra of weak radar echoes: application to Mars. 28: 301–306.

Lovelock, J. E., and Hitchcock, D. R. Life detection by atmospheric analysis. 7: 149–159.

Murray, B. C., Soderblom, L. A., Cutts, J. A., Sharp, R. P., Milton, D. J., and Leighton, R. B. Geological framework of the south polar region of Mars. 17: 328–345.

Mutch, T. A., Binder, A. B., Huck, F. O., Levinthal, E. C., Morris, E. C., Sagan, Carl, and Young, A. T. Imaging experiment: The Viking Lander. 16: 92–110.

Oyama, V. I. The gas exchange experiment for life detection: the Viking Mars lander. 16: 557–568.

Packer, E., Scher, S., and Sagan, C. Biological contamination of Mars. 2: 293–316.

Pieri, D. Distribution of small channels on the Martian surface. 27: 25–50.

Pollack, J. B., Veverka, J., Noland, M., Sagan, C., Hartmann, W. K., Duxbury, T. C., Born, G. H., Milton, D. J., and Smith, B.A. Mariner IX television observations of Phobos and Deimos. 17: 394–407.

Sagan, C. The long winter model of Martian biology. 15: 511–514.

Sagan, C., and Fox, P. The canals of Mars: an assessment after Mariner IX. 25: 602–612.

Sagan, C., and Pollack, J. B., Differential transmission of sunlight on Mars: biological implications. 21: 490–495.

Sagan, C., Veverka, J., Fox, P., Dubisch, R., Lederberg, J., Levinthal, E., Quan, L., Tucker, R., Pollack, J. B., and Smith, B. A. Variable features on Mars: preliminary Mariner IX television results. 17: 346–372.

Smith, J., and Born, G., Secular acceleration of Phobos and of Mars. 27: 51–54.

Veverka, J., Sagan, C., Quan, L., Tucker, R., and Eross, B., Variable features on Mars: comparison of Mariner 1969 and Mariner 1971 photography. 21: 317–368.

Science:

Dalgarno, A., and McElroy, M. B. Mars: is nitrogen present? 170: 167–168.

Farmer, C. B., and Davies, D. W. Viking: Mars atmospheric water vapor mapping experiment, preliminary report. 193: 776–780.

Hammond, A. L. Mars as an active planet: the view from Mariner IX. 175: 286–287.

Hargraves, R. B., Collinson, D. W., and Spitzer, C. R. Viking magnetic properties investigation: preliminary results. 194: 84–86.

Kieffer, H., et. al. Infrared thermal mapping of the Martian surface and atmosphere: first results. 193: 780–785.

Mutch, T. A., et. al. The surface of Mars: the view from lander I. 193: 791–800.

Mutch, T. A., et. al. Fine particles on Mars. 194: 87–91.

Nier, A. O., McElroy, M. B., and Yung, Y. L. Isotopic Composition of the Martian Atmosphere. 68–70.

Owen, T., and Biemann, K. Composition of the atmosphere at the surface of Mars: detection of Argon–36 and preliminary analysis. 193: 801–803.

Sagan, Carl, Lederberg, J., and Levinthal, E. C. Contamination of Mars. 159: 1191–1196.

Sagan, Carl, Toon, O. B., and Gierasch, P. J. Climactic change on Mars. 181: 1045–1049.

Shorthill, R. W., et. al. The soil of Mars (Viking I). 194: 91–97.

Soffen, G. A., and Snyder, C. W. The first Viking mission to Mars. 193: 759–765.

Tyler, G. L., Campbell, D. B., Downs, G. S., Green, R. R., and Moore, H. J. Radar characteristics of the Viking I landing site. 193: 812–815.

Scientific American:

Murray, Bruce C. Mars from Mariner IX. 228: 48.

Pollack, James A. Mars. 233: 81–91.

Sagan, Carl. The Solar System. 233: 23–31.

Sky and Telescope:

Beatty, J. K. December 1976. Vikings rest during conjunction. 404–409.

Staff Report. September 1976. New evidence favors big bang. 162–163.

Staff Report. September 1976. Mars viewed from the Viking I Orbiter. 171–174.

FICTION

Arnold, Edwin L. *Gulliver of Mars.* New York City: Ace Books.

Brown, Fredric. *Martians Go Home.* New York City: Ballantine Books, 1976.

——. *Rogue in Space.* New York City: Bantam Books, 1971.

Bradbury, Ray. *The Martian Chronicles.* New York City: Bantam Books, 1970.

Burroughs, Edgar Rice. *A Princess of Mars.* New York City: Ballantine Books, 1967.

——. *The Gods of Mars.* New York City: Ballantine Books, 1964.

——. *The Warlord of Mars.* New York City: Ballantine Books, 1963.

——. *Thuvia, Maid of Mars.* New York City: Ballantine Books, 1963.

——. *The Chessmen of Mars.* New York City: Ballantine Books, 1963.

——. *The Master Mind of Mars.* New York City: Ballantine Books, 1963.

——. *A Fighting Man of Mars.* New York City: Ballantine Books, 1964.

——. *Swords of Mars.* New York City: Ballantine Books, 1963.

——. *Synthetic Men of Mars.* New York City: Ballantine Books, 1967.

——. *Llana of Gathol.* New York City: Ballantine Books, 1963.

——. *John Carter of Mars.* New York City: Ballantine Books, 1965.

Compton, D. G., *Farewell Earth's Bliss.* New York City: Ace Books, 1971.

Heinlein, Robert A., *Podkayne of Mars.* New York City: Berkley Publishing, 1972.

——. *Stranger in a Strange Land.* New York City: Berkley Publishing, 1971.

Moorcock, Michael, *Warriors of Mars* (also known as *The City of the Beast*). New York City: Lancer Books, 1965.

————. *Blades of Mars* (also known as *The Lord of the Spiders*). New York City: Lancer Books, 1965.

————. *Barbarians of Mars* (also known as *The Masters of the Pit*). New York City: Lancer Books, 1965.

Spinrad, Norman. *Agents of Chaos*. New York City: Belmont-Tower Books, 1972.

Swift, Jonathan. *Gulliver's Travels*. New York City: Bantam Books, 1965.

Wells, H. G. *War of the Worlds*. New York City: Pocket Books, 1953.

Wyndham, John. *Planet Plane* (also known as *Stowaway to Mars*). New York City: Fawcett, 1972.

[NASA news kits, media releases, and press conferences supplied the bulk of the Mariner and Viking data—and the interpretations thereof—which were used in this book.

[The charter of the National Aeronautics and Space Administration makes it mandatory for that government organization to make available, on request, its publications, papers, and booklets. These cover the entire spectrum of space exploration, including all Mars-related studies. Information about this material, as well as the material itself, may be obtained by writing to the National Aeronautics and Space Administration, Washington D.C. 20546. The reader is advised to first procure NASA's *Aerospace Bibliography*, which lists nearly 700 books and pamphlets from NASA and other publishers that provide a foundation for both the amateur and professional study of outer space.

INDEX

(Note: page numbers in *italics* refer to illustrations)

240

Kooistra, J. A., 73
Kuiper, Gerard P., 74
Kummer, Jr., Frederic Arnold, 64

Labeled Release Experiment, *152,*
    155–56, 179–80, 182
Laffen, Patricia, 85
Lampland, C. O., 46, 70
Latham, Dr. Gary, 175–77
de Laurentiis, Dino, 88
Leighton, Robert, 116, 117–19
Leonov, Aleksi, 112
Levin, Dr. Gilbert, 181–82
Lieutenant Gulliver Jones: His Vacation
    (Arnold), 56–58
Lindbergh, Charles, 110
Lippershey, Hans, 28
Llana of Gathol (Burroughs), 61
Locked City (Ayre), 62–63, 67, 99
Longomontanus, 27
Louis XIV of France, 32
Lowell, Percival, *13,* 42–50, 51, *56,* 61, 62,
    65, 67, 69–70, 71, 73, 74, 79, 108, 124,
    152, 161, 200, 218, 222
Luna III, 113
Lundin, Vic, 87

McCartney, Paul, 212
McElroy, Dr. Michael, 169
McLaughlin, Dean B., 73–74
von Madler, J. H., *33*
Maedler, 35
Maid of Mars (Burroughs), 61
Mailer, Norman, 212
Mantee, Paul, 87, *97*
Marly, Florence, 88
Mariner program, the
    Mariner III, 114
    Mariner IV, *14,* 42, 67, *111,* 114–20
    Mariner VI, 120
    Mariner VII, 120
    Mariner VIII, 122
    Mariner IX, 74, *109,* 120, *121,* 122–23,
        124, *125, 127,* 128–30, *185,* 188
    Mariner X, *192*
Mars
    aphelions of, 44
    atmosphere of, 120, 128, 130, 144, *160,*
        164, 166–67, 168, 183–87
    basins on, 159
    biological experimentation on, 155–56,
        171, 177–87
    broadcasting from, *134*
    canals on, 42–43, 45, 46–50, 61, 67, 69,
        78, 106, 124
    color of, 77, 126–27, 166
    chemical composition of, 74–77, 126,
        149–51, 166–69, 177

craters on, *14,* 119, 120, 123, 128, *160,*
    169, 171, *184,* 188–91
deserts on, 46
diameter of, 44
distance between the earth and, 32
dust on, 71, 74, 126, *158, 182,* 187
dust storms on, 122–23, 124, 130
evolution of, 10, 126
experimentation on, 144, 149–53,
    155–56, 177–83, 187
gravity on, 61, 126
ionosphere of, 120
life on, 34–35, 42, 43, 47, 48–50, 70,
    71–77, 120, 152–56, 159, 167, 168–69,
    177–88, 220
in literature, 39–42, 50–68, 92–106
magnetic field of, 119, 144
maria of, 71, 73–74, 78, 122
mass of, 44
moons of, 36, 42, 44, 123, 124, *125, 129,*
    185, 187, 188, 191–93
movies and, *8,* 62, 72, 75, 76, 78–92, *93,*
    94, 98, 103, 190, 219
myths and, 37–39
orbit of, 27, 44
perihelions of, 44
phases of, 30
photographing of, 117–19, 120, 122–23,
    124, 127–30, 135, 146–49, 150, *153,*
    159, 161, 162, 166
photographs of, 7, *14, 111, 115, 118,*
    121, 154, 158, 160, 161, 162, 163, 164,
    165, 166, 167, 168, 169, 170, 172, 174,
    175, 180–81, 182, 183, 184, 185, 221
polar caps of, 35, 45, 61, 71, 74–77, 120,
    126, *184*
radioactive band of, 119
robot probes to, 7, *14,* 42, 67, *109, 111,*
    113, 114–24, *125, 127,* 128–30, 135–93.
rotation of, 35, 44, 45
science fiction and, 69, *189, 190,*
    193–205
seasons of, 45
Sinton's Bands, 71
soil of, *162, 163*
surface pressure of, 162–63, 173, 177
tectorism of, 127n
temperature on, 45, 74, 146, 163–64,
    173, 183
temperature gauging on, 151–52
terrain of, 45, 46, 47, 48, 106, 119, 120,
    128–30, 159, *163, 164,* 173–75, 177
tremors on, 151
troughs on, 128, *169*
volcanoes on, 127–28, 154, *161, 168*
weather on, 130, 173
weather gauging on, 151, *164*
winds on, 164, 173